ILLUMINATED MANUSCRIPTS

IN

OXFORD COLLEGE LIBRARIES,
THE UNIVERSITY ARCHIVES

AND THE

TAYLOR INSTITUTION

283 Christ Church 92, fol. 27. The King receives and repays gifts.

J. J. G. ALEXANDER AND ELŹBIETA TEMPLE

ILLUMINATED MANUSCRIPTS

IN

OXFORD COLLEGE LIBRARIES, THE UNIVERSITY ARCHIVES

AND THE

TAYLOR INSTITUTION

CLARENDON PRESS · OXFORD
1985

Oxford University Press, Walton Street, Oxford OX2 6DP
London New York Toronto
Delhi Bombay Calcutta Madras Karachi
Kuala Lumpur Singapore Hong Kong Tokyo
Nairobi Dar es Salaam Cape Town
Melbourne Auckland

and associated companies in
Beirut Berlin Ibadan Mexico City Nicosia

Oxford is a trade mark of Oxford University Press

Published in the United States
by Oxford University Press, New York

British Library Cataloguing in Publication Data
Alexander, J. J. G.
Illuminated manuscripts in the Oxford College libraries, the University archives and the Taylor Institution.
1. Illumination of books and manuscripts —Bibliography
I. Title II. Temple, Elżbieta.
III. Taylor Institution
011'.31 Z5948.M6
ISBN 0-19-817381-4

Library of Congress Cataloguing in Publication Data
Alexander, J. J. G. (Jonathan James Graham)
Illuminated manuscripts in the Oxford College libraries, the University Archives, and the Taylor Institution.
Includes Indexes.
1. Illuminated books and manuscripts—England—Oxford —Catalogs. I. Temple, Elżbieta. II. Title.
ND2894.G7A44 1984 745.6'7'09407402574 84-11216
ISBN 0-19-817381-4

Set by Eta Services (Typesetters) Ltd, Beccles, Suffolk
Printed in Great Britain
at the University Press, Oxford
by David Stanford
Printer to the University

PREFACE

THIS CATALOGUE of the illuminated manuscripts to be found in the libraries of the Oxford Colleges and two other institutions in Oxford, the University Archives and the Taylor Institution, has been compiled as a complement to three catalogues by Otto Pächt and J. J. G. Alexander published between 1966 and 1973. Those three catalogues listed the illuminated manuscripts of the Bodleian Library, Oxford, the first volume covering German, Dutch, Flemish, French and Spanish schools, the second the Italian schools and the third the British, Irish and Icelandic schools. A *Concordance of Bodleian Shelf-marks for volumes 1–3 and Addenda* by B. C. Barker-Benfield was published in 1974. The present catalogue, which lists manuscripts from all Western schools in a single volume, has been designed to the same format and compiled on the same principles. It aims to be a short-hand guide giving a summary indication of textual content, type of illumination, provenance and relevant bibliography with our attributions as to date and school for each manuscript. Though a number of reviewers of the earlier volumes and other users have regretted that more detailed information was not given as to the extent and content of the illumination, we are still persuaded that given our resources of time it was preferable to retain the principle of a summary guide in order to be able to publish the catalogue within a reasonable time span.

Even granted these more limited aims the project has encountered certain specific difficulties in its execution. First, obviously, is the geographical dispersal of many of the manuscripts in the various Colleges. Secondly, though some of the finest illuminated manuscripts in the College collections have been well known for a considerable time and have been studied in detail, the majority are relatively unexamined. Only two College collections have been catalogued more recently with the benefits of modern scholarship, Balliol College catalogued by Sir Roger Mynors in 1963, and Keble College by Mr. M. B. Parkes in 1979. For the majority of manuscripts the only printed catalogue is that by H. O. Coxe in two volumes of 1852. For the three catalogues of illuminated manuscripts in the Bodleian Library not only did the volumes of the *Summary Catalogue* and of the earlier 'Quarto' series of catalogues stand as it were behind each entry, but there was also a much more organized and systematic recording of corrections and bibliographical references, as well as a long tradition of scholarly expertise among the staff responsible for the collections within the Department of Western Manuscripts. How little many of the College manuscripts have been consulted or studied is indicated by the tiresome fact that so many of them have never been foliated in modern times. The reader should therefore be aware that there have been limits to the amount of time we have been able to spare to check such matters as the identification of the texts, the names of former owners or the existence of published references.

The College Libraries. Many of the Colleges, unlike the Bodleian Library, have retained a proportion of the manuscripts from their medieval libraries. The history of the formation and organization of these libraries as well as of their later preservation, dispersal or augmentation has been the object by now of a considerable amount of study. An overview with bibliography is given by P. Morgan, *Oxford Libraries outside the Bodleian. A Guide*, Oxford, 1974, 2nd edition 1980. Summary histories of most college libraries are also given in the *Victoria History of the County of*

Oxford, vol. 3, 1954. In addition the detailed scholarly catalogues of the manuscripts at Balliol and Keble Colleges already mentioned contain in their introductions accounts of the acquisition of the manuscripts of those libraries and of the main benefactors and collectors. For Merton College similarly there is F. M. Powicke's *The Medieval Books of Merton College*, 1931. For All Souls College there is *The History of All Souls College Library* by Sir Edmund Craster edited by E. F. Jacob, 1971, and N. R. Ker's *Records of All Souls College Library 1437–1600* (*Oxford Bibliographical Society Publications*, N.S. 16), 1971. On the New College Library there is an article by R. W. Hunt 'The medieval library' in *New College, Oxford 1379–1979*, eds. J. Buxton, P. Williams, 1979, pp. 317–45. Dr. Hunt's longer unfinished work on the history of New College Library is at present being prepared for publication by Dr. B. C. Barker-Benfield. Another paper by R. W. Hunt, 'The manuscript collection of University College, Oxford' is in *Bodleian Library Record*, 3, 1950, 13–34. The Jesus College library was discussed by T. J. Fordyce, T. M. Knox, 'The library of Jesus College, Oxford', *Oxford Bibliographical Society Proceedings*, V (1937), 53–115. For Lincoln College there are articles by R. Weiss, 'The earliest catalogues of the library of Lincoln College', *Bodleian Quarterly Record*, VIII (1937), 343–59, and by A. C. de la Mare, 'Vespasiano da Bisticci and the Florentine manuscripts of Robert Flemmyng in Lincoln College', *Lincoln College Record*, 1962–3, 12–14. V. Green, *The Commonwealth of Lincoln College 1427–1977*, 1979, also gives information about manuscript accessions. The history of the acquisition of humanistic manuscripts in the fifteenth century in Oxford is further treated by R. Weiss in his *Humanism in England during the fifteenth century*, 3rd edition, 1967, and in the Bodleian Library exhibition catalogue *Duke Humfrey and English Humanism in the fifteenth century* by R. W. Hunt and A. C. de la Mare, 1970.

The way the medieval College libraries in Oxford in general were built up, as also their contents and organization are discussed by N. R. Ker, 'Oxford College libraries before 1500', *The Universities in the late Middle Ages*, ed. J. Ijsewijn, J. Paquet, *Mediaevalia Lovaniensia*, VI (1978), 293–311. Evidence concerning the medieval book trade in Oxford is contained in G. Pollard, 'The University and the book trade in medieval Oxford', *Miscellanea medievalia*, III (1964), 336–44. Some information is given there about the 'pecia' system, for which see also G. Pollard, 'The *pecia* system in the medieval universities' in *Medieval scribes, manuscripts and libraries, essays presented to N. R. Ker*, eds. M. B. Parkes, A. G. Watson, 1978, pp. 145–61, and, when published, *La production des livres universitaires au moyen age: peccia et exemplar*, eds. L. Bataillou, R. H. Rouse. The system of the *cautio* by which books were pledged for loans is referred to by F. M. Powicke cited above and by G. Pollard, 'Mediaeval loan chests at Cambridge', *Bulletin of the Institute of Historical Research*, XVII (1939–40), 113–29. The chests are listed in *Statuta Antiqua Universitatis Oxoniensis*, ed. S. Gibson, 1931, index (s.v. 'chests'). N. R. Ker in his *Medieval Libraries of Great Britain*, 2nd edn. 1964, records the known surviving manuscripts which once belonged to but no longer remain in the College Libraries. He also gives valuable information on owners and donors and concerning medieval and later catalogues as well as more modern publications.

Much valuable information on the later history of the College libraries is contained in N. R. Ker's *Fragments of medieval manuscripts used as pastedowns in Oxford Bindings with a survey of Oxford Binding* c. *1515–1620*, 1954. Dr. Ker was also responsible for the catalogue *Oxford College Libraries in 1556. Guide to an exhibition held in 1956*, Bodleian Library, 1956. His Sandars lectures of 1955 'Oxford College Libraries in the sixteenth century' were published in the *Bodleian Library Record*, VI (1959), 459–515. J. N. L. Myres gives an account of 'Oxford Libraries in the seventeenth and eighteenth centuries' in *The English Library before 1700*, eds. F. Wormald, C. E. Wright, 1958. R. W. Hunt also wrote on St. John's College at this period, 'Donors of manuscripts to St. John's College, Oxford, during the presidency of William Laud, 1611–1621', *Studies in the Book Trade in honour of Graham Pollard*, (*Oxford Bibliographical Society*), 1975, 63–70. For Oxford bindings the reader should also consult S. Gibson, *Early Oxford Bindings* (*Bibliographical Society Illustrated Monographs* X), 1903, and J. B. Oldham, *English blind-stamped bindings*, 1952.

Illuminated Manuscripts in the libraries. From the point of view of the student of medieval book painting it will be apparent that the two largest groups of manuscripts here catalogued are, first, biblical commentaries, patristic works and scholastic texts of the thirteenth–fifteenth century, mainly English and French, and secondly texts connected with the new learning, classical, patristic and humanistic, these being mainly, but not exclusively, Italian fifteenth-century manuscripts. In other words these were study books which so far as we know were, in the main, bequeathed or given by College Fellows rather than bought by the Colleges. They are not the kind of manuscript which can be expected as a rule to be either illustrated or lavishly illuminated. Nevertheless not only are there some individual books of fine quality, but also as groups of texts they achieve great importance cumulatively as demonstrating types of book production with varieties of organizational practice in their lay-out and types of illumination. The study of such books is essential in order to understand the context within which grander illuminated manuscripts were produced. It is also worth stressing that the study of illumination can often provide students of texts with more exact information as to dating and provenance than is available by any other method.

The task of distinguishing English from French illumination in the thirteenth century has provided special difficulties which should be acknowledged. These are the outcome of a specific historical situation, for not only were French styles of illumination copied in England, but also, as we know from contemporary documents, many Englishmen worked as scribes, illuminators and even stationers in Paris and elsewhere (cf. no. 716, John Wade, an English scribe working in Paris). Clearly many 'University' books were also imported from continental universities. Though a number of books by or related in style to the work of William de Brailes, a known Oxford illuminator, have been found (nos. 190, 199, 200), a specifically Oxford style of illumination has not yet been identified in either the thirteenth or the fourteenth century. Such an identification must depend equally on palaeographical study. We can only hope that if it later proves possible the present catalogue may have facilitated the task by making more accessible some of the material.

The finest quality illumination in the Middle Ages is generally found in liturgical and devotional books and also in the later period in secular works made for particular royal or noble patrons. Such liturgical manuscripts as the Colleges possessed must have been mostly destroyed after the Reformation, as the fragments surviving as pastedowns in sixteenth-century Oxford bindings show. The Founder's Missal at Trinity College (no. 473), the All Souls Missal (no. 348), the University College Missal (no. 349) and Cardinal Wolsey's Gospel Lectionary and Epistolary at Magdalen and Christ Church respectively (nos. 828, 827) are exceptions, all having particular associations which must have helped to preserve them. The most richly illuminated manuscripts in the catalogue, therefore, manuscripts such as the Bohun Psalter from Exeter (no. 316), the Walter de Milmete from Christ Church (no. 283), the De Brailes Psalter from New College (no. 199) and the Amesbury Psalter from All Souls (no. 215) are all post-medieval acquisitions. One College library, that of Keble College, stands out in this respect as being quite a different kind of collection. Formed in response to the ideals of the Tractarian movement and with a specific interest in medieval liturgical practice, the library contains an exceptionally high proportion of fine illuminated continental manuscripts. Many of these come from the collection of Sir Thomas Brooke of Armitage Bridge, Huddersfield, which must be ranked as one of the more important of nineteenth-century English collections of illuminated manuscripts. They were bequeathed to the College by his brother. It is good to be able to point out that the collections are still growing as manuscripts are still given or bequeathed to Colleges, such as those given to Lincoln College by Mr.

Eric Sexton, 1972–3, those bequeathed by Lady Margaret Hall by Miss C. M. Borough in 1974 and most recently those given to New College by Sir John Nicholson in 1983.

A sad fact about the College manuscripts is that such a large number of them have been mutilated. We have made a distinction between mutilation and defacement of miniatures or initials. Some of the latter may have been done at the Reformation, e.g. the faces of the figures in the Crucifixion (no. 275, pl. XVII) or the Virgin's face in the Annunciation (no. 667, pl. XXXVIII) but some perhaps even before that. It is not clear when the mutilation which often takes the form of cutting out initials, took place, nor what the reasons for it were. A note in Queen's College MS. 349 (no. 818), to the effect that miniatures removed by an undergraduate were returned by the penitent thief in 1727, is important evidence that by that date mutilation was by would-be collectors. Samuel Pepys was given single leaves removed from early Insular manuscripts at Durham in 1700. Earlier than this it is possible that the objective of some thieves could have been the gold leaf backgrounds to initials, which might explain why, so far as is known, none of these mutilated pieces survive.

Aims and Scope of the lists. The main and very considerable work of checking all manuscripts to see whether or not they contained illumination was done by Dr. Temple who then wrote entries for them and did the necessary bibliographical research. Dr. Alexander has been through all the manuscripts with the preliminary descriptions, has assisted in dating, attribution and the selection of pages for reproduction, and is responsible for this Preface.

We have adopted similar rules to those observed in the Bodleian volumes, aiming to include all illuminated manuscripts up to 1600, other than heraldic or topographical works. We have not included printed books with manuscript illumination. We have included only a selection of manuscripts with penwork decoration where we felt it to be of exceptional quality or significant because dated or localized. We have continued to assign manuscripts to national schools according to the context and stylistic origin of their illumination, though aware of the difficulties and dangers of this. In a few cases where illumination of two schools occurs in the same manuscript it receives two entries (e.g. Balliol 2, nos, 700, 895). Medieval bindings have been noted.

Textual contents of the manuscripts. Description of the textual contents has intentionally been kept to a minimum. Fuller descriptions are to be found in H. O. Coxe, *Catalogus codicum MSS. qui in Collegiis Aulisque Oxoniensibus hodie adservantur*, 2 volumes 1852. Volume 1 (reprinted 1972) contains University, Balliol, Merton, Exeter, Oriel, Queen's, New College, Lincoln. Volume 2 contains All Souls, Magdalen, Brasenose, Corpus Christi, Trinity, St. John's, Jesus, Wadham, Worcester, Hertford. Christ Church manuscripts were catalogued by G. W. Kitchin, *Catalogus Codicum Manuscriptorum qui in Bibliotheca Aedis Christi apud Oxonienses adservantur*, 1867. Balliol manuscripts were re-catalogued by R. A. B. Mynors, *Catalogue of the manuscripts of Balliol College*, 1963.

The Keble College manuscripts were catalogued by M. B. Parkes, *The medieval manuscripts of Keble College*, Oxford, 1979. Manuscripts not included in any of these catalogues are described by N. R. Ker, *Medieval manuscripts in British Libraries*, vol. 3, 1983, pp. 582–735. Descriptions of recent accessions are also kept in Duke Humfrey's Library, the Bodleian Library.

All manuscripts are in Latin and written on parchment unless specified. In this volume we have given the dimensions of manuscripts.

Date and place of manufacture. Unless the evidence is specified, for example, colophons, liturgical evidence or ownership, the attribution and dating depend on the style of the illumination and script.

Bibliography. This has been limited intentionally and concentrates on works concerned specifically with illumination. Obvious reference works such as A. B. Emden, *A biographical Register of the University of Oxford to* A.D. *1500*, Oxford, 1957–9, *id.*, A.D. *1501–1540*, 1974, or R. Foster, *Alumni Oxonienses, 1500–1714*, 1891–2, *id.*, *1715–1886*, 1888, or N. R. Ker, *Medieval Libraries of Great Britain*, 1964, as also the catalogues listed above are omitted or quoted selectively, as citation would have involved very large numbers of manuscripts.

Provenance. In order to give a picture of the growth of the libraries we have recorded in this volume not only medieval but many of the later owners and donors.

Access. Though the manuscripts of a number of the colleges are at present housed in the Bodleian Library and can be ordered there in the normal way in Duke Humfrey's library, it will be advisable for intending readers to check with the Bodleian Library, since some colleges retain some of their manuscripts even when they deposit others, and also the situation may change. Colleges at present depositing manuscripts included in this volume are: Brasenose, Corpus Christi, Hertford, Jesus, Lady Margaret Hall, Lincoln, Magdalen, New College, Oriel, Trinity and University. In the case of manuscripts housed in the Colleges themselves, intending readers are strongly advised to write well in advance to the College Librarian concerned.

Acknowledgements. We should like first to thank all the Librarians of Colleges and their staffs who have not only made their manuscripts available to us, but have had the considerable task of transporting many of them to the Bodleian Library to be photographed. We should like also to thank all members of the Bodleian Library staff and especially those of Duke Humfrey's Library, who have facilitated our work. Dr. A. C. de la Mare of the Department of Western Manuscripts has been particularly generous with advice, help and information, especially, of course, concerning the Italian manuscripts. We have also benefited in numerous ways from help and information given by Dr. B. C. Barker-Benfield. They, as also Mr. P. Morgan and Mr. D. G. Vaisey, have read this Preface and thanks are due to them for various corrections and suggestions. It is a fine Bodleian tradition that there should be a mutual exchange of information between readers and staff and we also at various times have had the benefit of consulting visiting scholars to all of whom we extend our thanks.

The manuscript of this catalogue was expertly retyped by Mrs. M. Davies and prepared for the Press by Mrs. J. Dodsworth. The project was enthusiastically supported at its inception by the late Dr. R. W. Hunt, Keeper of Western Manuscripts, and the then Bodley's Librarian, Professor R. Shackleton. Since then gratitude is due to the present Keeper of Western Manuscripts, Mr. D. G. Vaisey, for his work in ensuring its continuation. The project was funded mainly by generous grants from the Hulme Surplus Fund with additional help from the Bodleian Library. Publication has been undertaken jointly by the Oxford University Press and the Bodleian Library.

CONTENTS

LIST OF ABBREVIATIONS

Anglo-Saxon illumination	*Anglo-Saxon illumination in Oxford libraries* (Bodleian Library Picture Book Special Series no. 1), introductory note by J. J. G. Alexander, Oxford, 1970.
Aristoteles Latinus I	G. Lacombe, *Aristoteles Latinus, Pars Prior* (Corpus Philosophorum Medii Aevi). Rome, 1939.
Boase, *English Art*	T. S. R. Boase, *English Art 1100–1216* (Oxford History of English Art, III), Oxford, rev. edn. 1971.
B.L.R.	*The Bodleian Library Record*, 1939–
B.Q.R.	*The Bodleian Quarterly Record*, 1914–38.
Branner	Robert Branner, *Manuscript Painting in Paris during the reign of Saint Louis*. University of California Press, Berkeley Los Angeles London, 1977.
Brieger, *English Art*	P. Brieger, *English Art 1216–1307* (Oxford History of English Art, IV), Oxford, 1957.
de la Mare, *A fragment*	A. C. de la Mare, 'A fragment of Augustine in the hand of Theodoricus Werken', *Transactions of the Cambridge Bibliographical Society*, VI (1976), 285–90.
de la Mare, *Vespasiano da Bisticci*	A. C. de la Mare, 'Vespasiano da Bisticci and the Florentine MSS of Robert Flemmyng in Lincoln College', *Lincoln College Record*, 1962–3, 7–16, pls. I–VI.
Destombes, *Mappemondes*	M. Destombes, *Mappemondes A.D. 1200–1300* (Imago Mundi, supplement no. IV), Amsterdam, 1964.
E.E.T.S., or. ser. and ex. ser.	*Early English Text Society*, original series and extra series.
Emden	A. B. Emden, *A Biographical Register of the University of Oxford to A.D. 1500*, 3 vols., Oxford, 1957–9.
Exh., *Archbishop Laud*	*Archbishop Laud 1573–1645*, Commemorative exhibition held in the Bodleian Library, Oxford, 1974.
Exh., *Duke Humfrey*	*Duke Humfrey and English Humanism in the fifteenth century*, Catalogue of an Exhibition held in the Bodleian Library, Oxford, 1970.
Exh., *Illuminated Manuscripts*, 1908.	Burlington Fine Arts Club, *Exhibition of Illuminated Manuscripts*, London, 1908.
Exh., *Italian Illuminated Manuscripts*	*Italian illuminated manuscripts from 1400 to 1550, Catalogue of an exhibition held in the Bodleian Library, Oxford, 1948* (catalogue entries by O. Pächt), Oxford, 1948.
Exh., *Manuscripts at Oxford*	*Manuscripts at Oxford: an exhibition in memory of Richard William Hunt (1908–1979), Keeper of Western Manuscripts at the Bodleian Library, Oxford, 1945–1975, on themes selected and described by some of his friends*. Edited by A. C. de la Mare & B. C. Barker-Benfield, Bodleian Library, Oxford, 1980.

Forshall and Madden	J. Forshall and F. Madden, *The Holy Bible containing the Old and New Testaments, with the Apocryphal Books, in the earliest English versions made from the Latin Vulgate by John Wycliffe and his followers*, Oxford, 1850.
Frere, I.	W. H. Frere, *Biblioteca Musica-Liturgica. A descriptive handlist of the musical and Latin-liturgical MSS. of the Middle Ages preserved in the Libraries of Great Britain and Ireland*. Vol. I, London, 1901.
H.B.S.	*Henry Bradshaw Society*.
Hunt, *Manuscript*	R. W. Hunt, 'The manuscript collection of University College, Oxford: origins and growth', *B.L.R.* III (1950), 13–34.
Hunt, *Medieval library*	R. W. Hunt, 'The medieval library', *New College Oxford 1379–1979*, eds. J. Buxton, P. Williams, Oxford, 1979, 317–45.
James, *Ancient Libraries*	M. R. James, *The Ancient Libraries of Canterbury and Dover*, Cambridge, 1903.
James, *Dr. John Dee*	*Lists of manuscripts formerly owned by Dr. John Dee* with preface and identifications by M. R. James (Supplement *Bibliographical Society Transactions*, 1) 1921.
J.W.C.I.	*Journal of the Warburg and Courtauld Institutes*, London.
Ker, *Catalogue*	N. R. Ker, *Catalogue of manuscripts containing Anglo-Saxon*, Oxford, 1960.
Ker, *College Libraries*	N. R. Ker, 'Oxford College libraries before 1500', *The Universities in the late Middle Ages*, ed. J. Ijsewijn, J. Paquet, *Mediaevalia Lovaniensia*, VI (1978), 293–311.
Ker, *Medieval Libraries*	N. R. Ker, *Medieval Libraries of Great Britain* (Royal Society, *Guides and Handbooks* n. 3, 2nd edn., London, 1964).
Ker, *Pastedowns*	N. R. Ker, *Fragments of medieval manuscripts used as pastedowns in Oxford bindings with a survey of Oxford Binding c. 1515–1620* (Oxford Bibliographical Society Publications, new series V, 1951–2), Oxford, 1954.
Millar, II	E. G. Millar, *English illuminated manuscripts of the XIVth and XVth centuries*, Paris and Brussels, 1928.
Morgan, *Early Gothic*	N. J. Morgan, *Early Gothic Manuscripts (1), 1192–1250* (Survey of Manuscripts illuminated in the British Isles, 4), 1982.
Morgan, *Gothic Mss.*	N. J. Morgan, 'Gothic manuscripts in England', *Liber Liborum. The Book through 5000 years*, ed. H. D. L. Vervliet, Brussels, 1972.
Mynors, *Catalogue*	*Catalogue of the manuscripts of Balliol College, Oxford*, compiled by R. A. B. Mynors, Oxford, 1963.
Mynors, *Werken*	R. A. B. Mynors, 'A Fifteenth-century scribe: T. Werken', *Transactions of the Cambridge Bibliographical Society*, I (1949–53), 97–104.
New Pal. Soc. I and II	*New Paleographical Society. Facsimiles of Ancient Manuscripts*, 1st and 2nd series, London, 1903–30.
Pächt, Alexander, 1, 2, 3	O. Pächt and J. J. G. Alexander, *Illuminated Manuscripts in the Bodleian Library, Oxford*, 3 vols., 1966–73.
Powicke, *Medieval Books*	F. M. Powicke, *Medieval Books of Merton College*, 1931.

(R)	Signifies a roll.
Randall, *Images*	L. M. C. Randall, *Images in the margins of Gothic manuscripts* (California Studies in the History of Art, IV), Berkeley and Los Angeles, 1966.
Rickert, *Miniatura* I, II	M. Rickert, *La miniatura inglese*, 2 vols., Milan, 1959–61.
Rickert, *Painting in Britain*	M. Rickert, *Painting in Britain. The Middle Ages* (The Pelican History of Art), 2nd edn., 1965.
Rural Life	*English rural life in the Middle Ages* (Bodleian Library Picture Book no. 14), introductory notes by J. W. Y. Higgs, Oxford, 1965.
St. Albans Psalter	*The St. Albans Psalter* (*Albani Psalter*), 1. *The full-page miniatures* by O. Pächt. 2. *The initials* by C. R. Dodwell. 3. *Preface and description of the manuscript* by F. Wormald, London, 1960.
Saunders, *English Illumination*	O. E. Saunders, *English illumination*, 2 vols., Florence and Paris, 1928.
Saxl and Meier, III	F. Saxl, H. Meier, *Verzeichnis astrologischer und mythologischer illustrierter Handschriften des lateinischen Mittelalters, III. Handschriften in englischen Bibliotheken* (Herausgegeben von H. Bober) 2 vols., London, 1953.
Swarzenski, *Monuments*	H. Swarzenski, *Monuments of Romanesque Art. The Art of Church Treasures in North-western Europe*, 2nd edn., London, 1974.
Thomson, *St. Albans*	R. M. Thomson, *Manuscripts from St. Albans Abbey 1066–1235*, Woodbridge, 1982.
Weiss, *Catalogues*	R. Weiss, 'The earliest catalogues of the Library of Lincoln College', *Bodleian Quarterly Record*, VIII (1935–8), 343–59.
Weiss, *Humanism*	R. Weiss, *Humanism in England during the fifteenth century* (Medium Aevum monographs, 4), 3rd edn., Oxford, 1967.

BRITISH MANUSCRIPTS

THE IMPORTANT examples of Anglo-Saxon illumination, the drawings in St. John's College MSS. 194 and 28 (nos. 1–2) and in the Gospels, Wadham College MS. 2 (no. 6), are already published. The latter illustrates, as does the early twelfth-century Life of St. Cuthbert from Durham, University College MS. 165 (no. 17), the continuity of Anglo-Saxon pictorial styles after the Conquest. The small unpublished initial in the Pseudo-Dionysius, St. John's College MS. 128 (no. 11), showing the *Christus super aspidem*, is also interesting in this respect.

Important twelfth-century monastic manuscripts in the collections are the Chronicle of Florence of Worcester, Corpus Christi College MS. 157 (no. 21), the Augustine from Evesham, Jesus College MS. 93 (no. 22) and the two leaves of the Deposition and the Three Maries at the Grave from St. Albans, Corpus Christi College MS. 2* (no. 25). There are also an important group of manuscripts from Cirencester (nos. 38–43) and examples of the typical Reading Abbey initial style (nos. 44, 52, 67). The provenance of other English monastic manuscripts, several with fine initials, remains to be established (nos. 36, 37, 46). The Missal from Bangor, Corpus Christi College MS. 122 (no. 23), shows the survival of Celtic styles in Wales into the twelfth century.

The glossed Isaiah, St. John's College MS. 26 (no. 65), is by an artist who worked for Abbot Simon of St. Albans (1167–83), and who was probably an itinerant professional. It therefore illustrates the beginning of a secular professional production which becomes the rule in the thirteenth century. A number of manuscripts also illustrate the so-called white-lion style of decoration, used especially in glossed books of the Bible and in the commentaries of Peter Lombard. This style, now thought to emanate from Paris and to be connected with the schools there, spread extremely quickly in all parts of Europe (nos. 104, 105, 138, etc.).

In the production of illuminated manuscripts in England in the thirteenth century, three texts stand out, Psalters, Bestiaries and Apocalypses. Magdalen College Lat. 100 (no. 171) is a Psalter with fine quality historiated initials some of which have been removed as they are on inserted pieces of parchment. St. John's College MS. 61 (no. 146) and St. John's College MS. 178 (no. 264) are examples of Bestiaries, and New College MS. 65 (no. 234), University College MS. 100 (no. 259) and Lincoln College MS. 16 (no. 282) of Apocalypses. The glossed Psalter, Merton College MS. 208 (no. 161) shows how interesting iconography can be found often in less important manuscripts. In this example an artist struggles to represent the Trinity. A Bible, Queen's College 52 (no. 158), is also interesting, in this case for an extremely strange style.

The thirteenth century is also well-known for the production of small one volume 'University' Bibles and naturally there are a great many in the collections. New College MS. 7 (173) is a good early example, interesting since it has written instructions in French and also preliminary marginal drawings. Other good examples are Merton College MS. 7 (no. 190) and Christ Church MS. 105 (no. 200), both from the workshop of William de Brailes who was also responsible for the well-known Psalter in New College, MS. 322 (no. 199). Other good Bibles are Jesus College MS. 55 (no. 222) made for Franciscan patrons, All Souls College MS. 2 (no. 228) and Corpus Christi College MS. 1 (no. 216), the last from the workshop of the illuminator of the Bible in the British Library written by William of Devon. New College MS. 20 (no. 205) is a Wisdom Books with gloss of fine quality, also interesting in having preliminary marginal drawings.

The very finest quality work of the mid-century is represented by the Amesbury Psalter, All Souls College MS. 6 (no. 215), which is well-known, but there are two other good Psalters of similar date, New College MS. 358 (no. 214) and Corpus Christi College MS. 285 (no. 217). The latter is an interesting example of a manuscript which has been updated later by the addition of borders and, it seems likely, some general retouching. The same happened to the Amesbury Psalter with the addition of the remarkable grotesques presumably commissioned by its owner in the fourteenth-century, John Grandison, bishop of

Exeter. A third example is yet another Psalter, Corpus Christi College MS. 17 (no. 371) where borders were added in England *c.* 1400 to a French thirteenth-century manuscript (no. 668). An exceptionally richly illuminated university book is the Aristotle, Merton College MS. 269 (no. 241), another example containing preliminary marginal drawings which have not been erased.

The lavish production of Psalters in the early fourteenth-century is represented by two manuscripts, Jesus College MS. 40 (no. 266) and All Souls College MS. 7 (no. 274), the former perhaps from London, the latter perhaps from Ely. The manuscripts of this period are famous for their marginal drolleries and in the Jesus College Psalter even the resurrected dead are shown as nude figures wrestling pick-a-back. A fine quality coloured drawing in the University Chancellor's Book, University Archives A.1 (no. 275), though unfortunately defaced, is important as probably done in Oxford. Other books of the period *c.* 1320–40 close in style to hands found in the well-known Milmete manuscript, Christ Church MS. 92 (no. 283), are linked some with London (no. 280), some with Oxford (no. 281).

Of the later fourteenth-century manuscripts the most important is the Bohun Psalter, Exeter College MS. 47 (no. 316). Another Psalter of somewhat earlier date is less well-known, Exeter MS. 46 (no. 306), and is an example of the expressive mid-century style. The Chancellor's Book (no. 317) also has a representation of Edward III dated 1375 which bears a close similarity to the Westminster Abbey effigy of the King.

Two important Missals of the late fourteenth century are the Lapworth Missal dated 1398, Corpus Christi College MS. 394 (no. 333) and Trinity College MS. 8 (no. 332). Keble College MS. 47 (no. 361) is a lavish Book of Hours with illumination related to that found in the Carmelite Missal, whilst Trinity College MS. 77 (no. 362), a Sarum Lectionary not previously illustrated, relates to illumination in the Lytlington Missal. Oriel College MS. 75 (no. 400) of the first quarter of the fifteenth-century, is another Missal whose fine Canon page has not been reproduced before. Two copies of Gower's *Confessio Amantis*, Corpus Christi College MS. 67 (no. 391) and All Souls College MS. 98 (no. 392) have illumination close in style to Hermann Scheerre. A Book of Hours, St. John's College MS. 94 (no. 418) is an interesting manuscript written and illuminated with great lavishness in a crude but expressive style by John Lacy, a Dominican recluse at Newcastle-on-Tyne. Presumably he was a self-taught amateur illuminator. All Souls College, MS. 10 (no. 462) containing the French translation of the Gospels by Guyart Desmoulins, and Merton College MS. 319 (no. 472), containing Thomas Netter of Walden's Doctrinale Fidei, which he wrote *c.* 1421–6, are good examples of English illumination of the second quarter of the fifteenth-century. The latter has a representation of the King at Mass and below this a scene of the burning of heretical books, no doubt Lollard texts. A group of books in the style of William Abell, the London illuminator, include a Gower *Confessio Amantis*, New College MS. 266 (no. 492), medical texts, St. John's MS. 18 (no. 495), Margaret of Anjou's Prayer Roll, Jesus College MS. 124 (no. 564) and the winter part of the Abingdon Missal, dated 1461, Trinity College MS. 75 (no. 578).

By this time in the mid 1450's Italian humanistic decorated manuscripts had reached England and it is interesting to see English artists evolving English forms of white-vine initials for texts written in forms of humanistic script for English humanist patrons, for example Balliol College MS. 131 (no. 562), New College MS. 228 (no. 569), Balliol College MS. 286 (no. 571) or Balliol College MS. 290 (no. 614). The Chaundler manuscript, New College MS. 288 (no. 581), of *c.* 1463 looks to Flemish rather than to Italian stylistic models, however, in its remarkable topographical representations of the college.

An interesting form of book of the later fifteenth century is the Genealogical Chronicle of the Kings of England starting from the Creation which is constructed of a roll bound up in codex form. Several examples exist in the collections (nos. 585, 602, 624) and Jesus College MS. 114 (no. 587) has a drawing of Edward IV and his queen, Elizabeth Woodville, adoring the Trinity. Another common text of the period often illuminated with royal portraits is the Statutes of the Realm, represented by an earlier example of before 1435, Merton College MS. 297B (no. 436) and by St. John's College MS. 257 (no. 621) of the last years of the fifteenth century.

Finally it should be noted that there are a number of manuscripts with drawings, some added on flyleaves (nos. 329, 338) some as decorations to catchwords (nos. 408, 416, 419, 527, 570) and some as illustrations, Corpus Christi College MS. 161 (no. 405), a *Speculum humanae salvationis*, University College MS. 82 (no. 548), the Cartulary of Beverley Minister, and Corpus Christi College MS. 78 (no. 368) a Brut Chronicle.

(*Measurements, e.g.* (*164 × 127*) *are given in millimetres*)

1. **St. John's College 194,** fol. 1^{v} *c.* 950
Christ Church, Canterbury (?)

Gospels (164 × 127)

Fine minature (partly coloured drawing) added on a fly-leaf of a Breton (?) MS. of saec. IX/X, retouched, possibly in saec. XV. Faint drawing fol. 31^{v}, saec. X (?). Cf. no. 508. [Illus.]

Prov. Pressmark 'GL', Christ Church, Canterbury, saec. XII (?), fol. 1. Bequeathed by Henry Price, fellow (d. 1600), fol. 3.

James, *Ancient Libraries*, App. D, 527; F. Wormald, *English Drawings of the tenth and eleventh centuries*, 1952, 23, 77, no. 52; pl. 40b; *Anglo-Saxon Illumination*, pl. 5; E. Temple, *Anglo-Saxon Manuscripts 900–1066*, 1976, cat. 12, ill. 47.

2. **St. John's College 28,** fols. 1–4^{v}, 78–81^{v}
Saec. X med.
St. Augustine's, Canterbury (?)

Miscellaneous texts (328 × 257)

Important drawing on the original flyleaf, fol. 2. Fine sketches, nearly contemporary, added on the end leaf, fol. 81^{v}. [Illus.]

Prov. Bound up with a MS. written in Canterbury, cf. no. 4. Possibly at Abingdon in saec. XII in, (St. Vincent invoked, fol. 81). Given by sir John White of Southwick, Hants., 1553. May, therefore, have come from Southwick Priory like other of his gifts.

F. Wormald, *English Drawings of the tenth and eleventh centuries*, 1952, 25, 37, 63, 74, 77–8, pl. 2; B. Raw, 'The drawings of an angel in MS. 28, St. John's College, Oxford', *J.W.C.I.*, XVIII (1955), 318–19; Ker, *Catalogue*, no. 361; *Anglo-Saxon Illumination*, pls. 7, 8; E. Temple, *Anglo-Saxon Manuscripts 900–1066*, 1976, cat. 13, ills. 42, 43; O. Pächt, 'Drawings on fly-leaves', Exh., *Manuscripts at Oxford*, VII. 1; J. E. Rosenthal, 'Three drawings in an Anglo-Saxon Pontifical: Anthropomorphic Trinity or a Threefold Christ?', *Art Bulletin*, LXIII (1981), 552 n. 27, fig. 8.

3. **Oriel College 3** Saec. X ex.
Christ Church, Canterbury

Prudentius, Peristephanon, Cathemerinon, etc. (317 × 205)

Fine initials. [Illus.]

Prov. Bequeathed by John Tailour, fellow, provost 1479–92, fol. 5.

Emden, 1851; Ker, *Catalogue*, no. 358; *Anglo-Saxon Illumination*, pl. 14c; T. A. M. Bishop, 'Notes on Cambridge Manuscripts, Part VII', *Cambridge Bibliographical Society Transactions*, V (1963), 415–16, 420, 421 (25); E. Temple, *Anglo-Saxon Manuscripts 900–1066*, 1976, cat. 19 (viii), ills. 71, 72.

4. **St. John's College 28,** fols. 7–77^{v} *c.* 1000
St. Augustine's, Canterbury

Gregory the Great, Pastoral Care (330 × 240)

Fine initials. The scribe also wrote Cambridge, MS. C.C.C. 389, *c.* 1000, of St. Augustine's, Canterbury, provenance. Cf. no. 2.

T. A. M. Bishop, *Codex Leidensis Scaligeranus 69* (Umbrae Codicum Occidentalium, X), 1966 (25); *Anglo-Saxon Illumination*, pls. 14a, b; T. A. M. Bishop, *English Caroline Minuscule*, 1971, 3, pl. 5; Exh., *English illuminated manuscript 700–1500*, Bibliothèque Royale, Brussels, 1973, no. 3; E. Temple, *Anglo-Saxon Manuscripts 900–1066*, 1976, cat. 13.

5. **St. John's College 17** *c.* 1090
Ramsey or Thorney

Computistical tracts. In Latin and Old English (340 × 255)

Fine drawings, *mappa mundi*, diagrams, tables, initials. [Illus.]

Medieval boards with later leather cover.

Prov. Thorney abbey, *c.* 1100: annals added in a nearly contemporary hand under abbot Gunter (d. 1092) and up to 1110, fols. 139–43. Annotations by Thomas Gascoigne, chancellor of Oxford University (d. 1458), fols. 72^{v}, 73^{v}. Robert Talbot, antiquary (d. 1558). Given by Hugh Wickstead, merchant taylor, saec. XVII in. Borrowed by Robert Cotton (d. 1631), who detached 5 leaves now forming fols. 80–4 of Brit. Libr. MS. Cotton Nero C. VII.

C. and D. Singer, 'Byrhtferd's Diagram', *B.Q.R.*, II (1917), 47–51, fig. 3; N. R. Ker, 'Membra disiecta', *British Museum Quarterly*, XII (1938), 131; id., *Catalogue*, no. 360; H. Bober, '"In Principio", Creation before Time', *Essays in honour of E. Panofsky. De Artibus Opuscula, ed.* M. Meiss, 1961, II, 21, pl. 7, fig. 6; *id.*, 'An illustrated Mediaeval School-Book of Beda 'De natura Rerum'', *Journal of the Walters Art Gallery*, XIX–XX (1956–7), 77 n. 32, 78, 87, figs. 9, 11–12, 15, 18; Destombes, *Mappemondes*, no. 25.8; R. W. Southern, 'The place of England in the 12th-century Renaissance', *Medieval Humanism and Other Studies*, 1970, 164 ff., pls. 3, 5, 6; C. Hart, 'The Ramsey Computus', *English Historical Review*, LXXXV (1970), 29–44; C. M. Kauffmann, *Romanesque Manuscripts 1066–1190*, 1975, cat. 9, ill. 21, fig. 41; M. Lapidge 'The revival of Latin learning in late Anglo-Saxon England', Exh., *Manuscripts at Oxford*, IV. 5; A. D. v. den Brinken, 'Raum und Zeit in der Geschichtsenzyklopädie des hohen Mittelalters', *Beiträge zur Geschichte von Stadt und Stift Essen*, Heft 96 (1981), 17 n. 38, Abb. 9.

6. **Wadham College 2** (A.10.22) Saec. XI ex.
Gospels (224 × 130)

Good miniatures (outline drawings), borders. Faint pencil sketch, fol. 13. Good historiated and other initials. [Illus.]

Prov. Given in 1625 by William Boswell, fellow, fol 2.

Exh., *Illuminated Manuscripts*, 1908, 7, no. 15, pl. 21; D. Talbot Rice, *English Art 871–1100*, 1952, 200, 209, pls. 64a, 73b; *Anglo-Saxon Illumination*, pls. 31–4; C. M. Kauffmann, *Romanesque Manuscripts 1066–1190*, 1975, Cat. 5, ills. 17, 18.

7. **Jesus College 37** Saec. XI ex.
Johannes Levita, Life of St. Gregory (243 × 180)

Good initials. [Illus.]
Medieval binding of white leather on boards, repaired.

Prov. Pressmark of St. Guthlac's priory, Hereford, 'XXus IIIIus', saec. XII–XIII, fol. 2. Bequeathed by sir John Prise (d. 1555).

N. R. Ker, 'Medieval pressmarks of St. Guthlac's Priory (Hereford) and Roche Abbey, Yorks.', *Medium Aevum*, v (1936), 47–8; *id.*, 'Sir John Prise', *The Library*, x (1955), 14 (J. 42).

8. **Exeter College 4** *c.* 1100
English or Norman
Priscian, Institutiones grammaticae (251 × 158)

Fine historiated and other initials, some partly over-painted later (unfinished). Penwork initials added *c.* 1200, fols. 118, 124, 132ᵛ. [Illus.]

Prov. William of Kirkeby, possibly rector of St. Mildred's, Oxford (d. 1398), and others, saec. XIV med., fol. 142ᵛ.

M. T. Gibson, 'Priscian, 'Institutiones Grammaticae': a handlist of manuscripts', *Scriptorium*, XXVI (1972), 117; *id.*, 'Pre-scholastic learning', Exh., *Manuscripts at Oxford*, 46, IX. 4.

9. **Balliol College 273** Saec. XII in.
Cicero, de inventione (226 × 147)

Good initials (unfinished). Scribbles and rough drawing, fols. 26, 26ᵛ, saec. XV (?).

Prov. Given in 1813 by the future sir William Hamilton, the philosopher (d. 1856), fol. ii.

10. **Lincoln College Lat. 27,** fols. 85–180 Saec. XII in.
Macrobius, Comm. in Somnium Scipionis (220 × 125)

Diagrams, *Mappa mundi*, fol. 160ᵛ. [Illus.]

Prov. Cf. no. 125.

Destombes, *Mappemondes*, no. 21, 30.

11. **St. John's College 128** Saec. XII in.
Dionysius Areopagita, de coelestia hierarchia, with gloss (225 × 130)

Good historiated initial. Other initials (unfinished). Diagrams. [Illus.]

Binding of blind-stamped leather on boards by Thomas Uffington (1477–96), Oxford (cf. J. B. Oldham, *English blind-stamped bindings*, 1952, pl. XVIII, 167, 173).

Prov. John Marsham, commoner, 1621, fol. 2.

G. Pollard, 'The names of some English fifteenth-century binders', *The Library*, XXV (1970), 210.

12. **University College 87** Saec. XII$\frac{1}{4}$
Augustine, contra Faustum (304 × 205)

Fine and other initials. [Illus.]

13. **Balliol College 307** Saec. XII$\frac{1}{4}$
Theological miscellanea (290 × 170)

Fine initial. Other initials. [Illus.]

Prov. Cluniac priory of St. Andrew, Northampton, pressmark 'D.O.', fol. 1; *ex libris* (erased), fol. 4.

14. **Balliol College 173A,** fols. 75–119ᵛ Saec. XII$\frac{1}{4}$
Ps.-Jerome, Epistle to Dardanus, etc. De generibus musicae (223 × 140)

Initials. Fine diagrams (partly coloured). [Illus.]

Prov. William Gray, bp. of Ely 1454–78 (?).

C. Page, 'Biblical instruments in medieval manuscript illustration', *Early music*, v (1977), 303 n. 25, pls. 9–11.

15. **Magdalen College Lat. 26** Saec. XII$\frac{1}{4}$
Liber de sacramentis ecclesie (160 × 92)

Fine initial.

Prov. Given in 1473 in memory of sir John Fastolf (d. 1459) by William Worcester, his secretary (d. 1480–83), to the founder, William Waynflete, bp. of Winchester 1447–86, fols. iᵛ, 1.

Emden, 2003, 2087.

16. **St. John's College 88** Saec. XII$\frac{1}{4}$
Rhabanus Maurus, de universo (250 × 160)

Arabesque initials.

Prov. Pressmark 'IX' of Chichester cathedral, fol. 1, saec. XIV.

17. **University College 165** *c.* 1120
Durham

Bede, Life of St. Cuthbert (195 × 125)

Important miniatures (one painted, the rest in coloured outline). Fine historiated and other initials. Preliminary marginal sketches. By the same artist as Durham

Cathedral Library MSS. Hunter 100 and B. IV. 14, part (c) and Bodl. MS. Digby 20, fol. 194. The scribe also wrote Durham Cathedral Library MS. B. IV. 14 (part b, fols. 112–69) and part of Hunter 100. [Illus.]

Prov. Ex libris (largely erased) of Southwick priory, Hants., saec. XII/XIII, fols. 1^{v}–2. William Roger of Painswick, commoner, 1666, first pastedown, pp. iii. 166, 200.

New Pal. Soc., II, pl. 67; B. Colgrave, 'The St. Cuthbert Paintings on the Carlisle Cathedral Stalls', *Burlington Magazine*, LXXIII (1938), 21; F. Wormald, 'Some illustrated MSS. of the Lives of Saints', *Bulletin of the John Rylands Library*, XXXV (1952–53), 259; Boase, *English Art*, 28, pl. 6c; *English Romanesque illumination* (Bodleian Library Picture Book no. 1) 1951, pl. 6; 'The medieval home of the illustrated life of St. Cuthbert', *B.L.R.* V (1954), 6; Hunt, *Manuscripts*, 21; O. Pächt, *The Rise of Pictorial Narrative in 12th-century England*, 1961, 14–18, 20 f., figs. 2–3; Rickert, *Painting in Britain*, 63, 228 n. 14, 233 n. 16, pl. 59b; J. J. G. Alexander, *Normal Illumination at Mont St. Michel 966–1100*, 1970, 141, pl. 34d; C. M. Kauffmann, *Romanesque Manuscripts 1066–1190*, 1975, cat. 26, ills. 59–63; E. Temple, 'A note on the University College Life of St. Cuthbert', *B.L.R.*, IX (1978), 320–2; M. Baker, 'Medieval illustrations of Bede's *Life of St. Cuthbert*', *J.W.C.I.*, XLI (1978), 16–49, pls. 1d, 3c, d, 5c, d, 6b, c, d, e, 8a, b, 9g; A. Lawrence, 'The influence of Canterbury on the collection and production of manuscripts at Durham in the Anglo-Norman period', *The Vanishing Past. Studies of Medieval Art, Liturgy and Metrology presented to Christopher Hohler, eds.* A. Borg, A. Martindale, 1981, 100, pls. 8, 34; K. Csapodi-Gardonyi, *Europäische Buchmalerei*, 1981, 54, plate.

18. **Oriel College 2,** fols. 1–184 — *c.* 1120–30, English (?)

Isidore, Etymologiae (350 × 220)

Fine initials (unfinished, mutilated). [Illus.]

Original white skin cover and pastedowns preserved as fly-leaves.

Prov. Possibly bequeathed by Simon Bredon, fellow of Merton College, archdeacon of Chichester (d. 1372).

Emden, 258.

19. **Christ Church 115** — *c.* 1120–30, St. Albans

Peter Damian, Augustine, etc. (245 × 160)

Fine initials close in style to Brit. Libr. MS. Royal 13 D. VI–VII from St. Albans. [Illus.]

Contemporary binding of white leather on boards.

Prov. Ex libris of St. Albans, saec. XII ex., fol. 1. Samuel Purchas, M.A., rector of St. Martin's. Ludgate (d. *c.* 1628), fol. ii.

Boase, *English Art*, 98–9; *St. Albans Psalter*, 150, 168, pls. 152f, 153g; N. R. Ker, *English Manuscripts in the century after the Norman Conquest*, 1960, pl. 21b; R. W. Hunt, 'The library of St. Albans', *Medieval Scribes, Manuscripts and Libraries. Essays presented to N. R. Ker, eds.* M. B. Parkes, A. G. Watson, 1978, 258 n. 35, 261; Thomson, *St. Albans*, 23–4, 111–2, pls. 32–3, 56–8.

20. **Trinity College 39** — *c.* 1130, Lanthony

Gregory the Great, Moralia in Job (305 × 180)

Good initials. Marginal drawings by the scribe. Later marginal sketch, fol. 18.

Prov. Note of contents written by Morgan of Carmarthen, canon of Lanthony, saec. XV, f. ii. Given in 1633 by Francis Baber, fellow, chancellor of the diocese of Gloucester (d. 1669), fol. 1.

Ker, *Medieval Libraries*, 108 n. 5, 111 n. 6, 112; E. G. W. Bill, *A Catalogue of Manuscripts in Lambeth Palace Library*, I, 1972, 11 n. 1; N. R. Ker, 'The English Manuscripts of the Moralia of Gregory the Great', *Kunsthistorische Forschungen Otto Pächt zu ehren, eds.* A. Rosenauer, G. Weber, 1972, no. 22.

21. **Corpus Christi College 157** — *c.* 1130–40, Worcester

Chronicle of Florence and John of Worcester (320 × 243)

Important miniature (tinted drawing). Fine miniatures (coloured drawings) added *c.* 1140, fols. 382–3. Companion volume to Bodl. MS. Auct. F.I.9, written mainly before 1130 by the same hand, probably John of Worcester, author of the Chronicle from 1118 till its end in 1140. [Illus.]

Binding of brown leather on original boards.

Prov. Exchanged in 1489 by frater Thomas Straynsham with Thomas Powycke, monk of Great Malvern, for a copy of Guido da Columna, *De Bello Troiano*, fol. i. Given by Henry Parry, fellow, 1613, fol. 1.

New Pal Soc., II, pls. 86–7; F. Wormald, 'The Survival of Anglo-Saxon illumination after the Norman Conquest', *Proceedings of the British Academy*, XXX (1944), 141 f., pl. 7; Boase, *English Art*, 157, 165; J. E. Hunt, *English and Welsh Crucifixes*, 1956, 38, 40, fig. 16; *Rural Life*, pl. 17a; Ker, *Medieval Libraries*, 129, 209, 281, 319; Rickert, *Painting in Britain*, 64, pl. 60; Exh., *The Bodleian Library. Heraldry*, 1967, 14, pl. 1; R. B. Green, 'A Typological Crucifixion', *Festschrift Ulrich Middeldorf, eds.* A. Kosegarten, P. Tigler, 1968, 138, pl. XII; G. Schiller, *Iconography of Christian Art*, II, 1971, 126, fig. 434; L. Wehrhahn-Stauch, 'Christliche Fischsymbolik', *Zeitschrift für Kunstgeschichte*, XXXV (1972), 39, fig. 45; Exh., *English Illuminated Manuscripts*, Brussels, 1973, no. 28, pl. 13; Swarzenski, *Monuments*, fig. 279; L. M. Ayres, 'The

Role of an Angevin Style in English Romanesque Painting', *Zeitschrift für Kunstgeschichte*, XXXVI (1974), 209, fig. 25; C. M. Kauffmann, *Romanesque Manuscripts 1066–1190*, 1975, cat. 55, ills. 139, 141–2.

22. **Jesus College 93** *c.* 1140
Evesham

Augustine, Comm. on Pss. I–C (385 × 190)

Fine historiated initials. Fine arabesque initials. [Illus.]

Contemporary binding of white leather on boards, pin and strap, bookmarker, tabs.

Prov. Evesham abbey, Worcs., (evidence from binding). Sir John Prise (d. 1555).

Boase, *English Art*, 47, n. 1; N. R. Ker, 'Chaining from a staple on the back cover', *B.L.R.*, III (1950–51), 104; *id.*, 'Sir John Prise', *The Library*, X (1955), 14, 19 (J.34); *id.*, *English Manuscripts in the century after the Norman Conquest*, 1960, 41; *id.*, *Medieval Libraries*, x, 81; G. Pollard, 'The construction of English twelfth-century bindings', *The Library*, XVII (1962), 16; A. Heimann, 'A twelfth-century manuscript from Winchcomb and its illustrations', *J.W.C.I.*, XXVIII (1965), 109.

23. **Corpus Christi College 122** *c.* 1140
Bangor (?)

Gospels (215 × 185)

Fine initials. The game or playing-board of the Gospels, *Alea Evangelii*, fol. 5. [Illus.]

Prov. Given by Henry Parry, fellow, 1619, fol. i.

J. A. Robinson, *The Times of St. Dunstan*, 1923, 7, 69, 171 ff.; F. Henry, G. L. Marsh-Micheli, 'A century of Irish illumination (1070–1170)', *Proceedings of the Royal Irish Academy*, LXII (1961–3), 101–2, 111, 152–5, pls. XXVIII–XXXII, XLIII; F. Henry, *Irish Art in the Romanesque Period*, 1970, 68.

24. **St. John's College 20** *c.* 1140
Haimo, Comm. on Isaiah (345 × 238)

Good historiated and other initials. Marginal drawing, fol. 35, [Illus.]

Prov. Cluniac Priory of St. Andrew, Northampton, letter-mark 'B.g.', fol. 1. Richard Butler, archdeacon of Northampton (d. 1612); given through the good offices of John Buckeridge, bp. of Rochester and Ely (d. 1631), in 1613, fol. 2.

25. **Corpus Christi College 2*** *c.* 1140–50
St. Albans

Two leaves perhaps from a Psalter (347 × 241)

Two important miniatures (unfinished). By the same hand as the calendar pictures of Bodl. MS. Auct. D.2.6 (cf. Pächt, Alexander, 3, no. 117), possibly early work of the Master of the Lambeth Bible. [Illus.]

Prov. On the reverse (fols. 2^{v} and 1^{r}) rough notes and part of a map of Palestine by Matthew Paris, *c.* 1250. The bifolium was formerly bound into MS. Corpus Christi College 2. Cf. no. 201.

M. R. James, *La estoire de Saint Aedward le Rei* (Roxburghe Club), 1920, 29, 73–4; Boase, *English Art*, 178; *St. Albans Psalter*, 158 n. 3, pl. 157e; R. Vaughan, *Matthew Paris*, 1958, 245 f., pl. 17; C. M. Kauffmann, *Romanesque Manuscripts 1066–1190*, 1975, 101–2, cat. 72. ills. 196–7; R. M. Thomson, *Manuscripts from St. Albans Abbey 1066–1235*, 1982, 31, no. 59, pls. 117–18.

26. **Queen's College 319,** fols. 43–128 Saec. XII$^{2}_{4}$
Hugh of St. Victor, de sacramentis (277 × 180)

Fine initial. [Illus.]

Binding of original boards with later velvet covering, metal bosses and clasps.

27. **New College 54** Saec. XII$^{2}_{4}$
English (?)

Gospel of St. John with gloss (220 × 160)

One good arabesque and other initials.

Prov. 'Liber Roberti presbiteri', fols. i^{v}, 1; 'precii XIII s., IIII d', fol. 1. Given by the founder, William of Wykeham, bp. of Winchester, 1367–1404, fol. i^{v}.

28. **Jesus College 69** Saec. XII$^{2}_{4}$
Pershore (?)

Bede, Comm. on the Canonical Epistles (303 × 205)

Fine historiated initial. Arabesque initials. [Illus.]

Contemporary binding of white leather on boards, repaired. Book-marker.

Prov. Possibly from Pershore abbey, Worcs. Sir John Prise (d. 1555).

N. R. Ker, 'Sir John Prise', *The Library*, X (1955), 13 (J. 16 ?); Pächt, Alexander, 3, no. 151.

29. **Jesus College 50** Saec. XII$^{2}_{4}$
West Country (?)

Ivo of Chartres, Panormia (pars altera) (260 × 170)

Good initials (unfinished). [Illus.]

Contemporary binding of white leather on boards, repaired.

Prov. Sir John Prise (d. 1555).

N. R. Ker, 'Sir John Prise', *The Library*, X (1955), 13 (J. 5).

30. **Jesus College 43** Saec. XII$^{2}_{4}$
Gloucester

Ps.-Athanasius, etc. (mutilated) (240 × 160)

Good arabesque initials.

Medieval binding of white leather on boards, repaired.

Prov. St. Peter's abbey, Gloucester, (table of contents, saec. XII, fol. 1).

Prov. Sir John Prise (d. 1555).

N. R. Ker, 'Sir John Prise', *The Library*, X (1955), 13, 18 (J. 8).

31. **Jesus College 64** Saec. XII^2_4
Evesham (?)

Bede, Comm. on the Song of Songs (283 × 185)

Fine initial. Other good initials.

Contemporary boards with white leather cover, tabs, repaired.

Prov. Contemporary letter of the abbot of St. Albans to the abbots of Evesham and Tewkesbury, fols. 1–2^{v}. Sir John Prise (d. 1555).

N. R. Ker, 'Sir John Prise', *The Library*, X (1955), 14, 20 (J. 45).

32. **Jesus College 102** Saec. XII^2_4
Winchcomb

Augustine, Comm. on Pss. LI–C (mutilated) (452 × 305)

Good arabesque initials. [Illus.]

Original boards with remains of blind-stamped leather binding by the 'Fishtail binder', probably Christopher Coke, Oxford, *c.* 1495; marks of a label on the back cover.

Prov. Corrections of saec. XII by the same hand as in Bodl. MS. Douce 368 from Winchcomb. Notes by Sir John Prise (d. 1555).

N. R. Ker, 'Sir John Prise', *The Library*, X (1955), 14, 19–20 (J. 35); *id.*, *English Manuscripts in the century after the Norman Conquest*, 1960, 7, 41, 50, pl. 25(c); A. Heimann, 'A twelfth-century manuscript from Winchcomb and its illustrations', *J.W.C.I.*, XXVIII (1965), 107 n. 74; G. Pollard, 'The names of some English fifteenth-century binders', *The Library*, XXV (1970), 197 n. 2, 210; N. R. Ker, 'The correction of mistakes in twelfth-century manuscripts, illustrated from Winchcombe books', Exh., *Manuscripts at Oxford*, VI. 2, fig. 20.

33. **Balliol College 178A** Saec. XII^2_4
Woburn

Florus diaconus, Comm. on St. Paul's Epistles. Part I (400 × 275)

Arabesque initials. Companion volume to MS. 178B (cf. no. 118).

Prov. Contemporary *ex libris* of Woburn abbey, Beds., fol. 1^{v}. William Gray, bp. of Ely 1454–78, fol. 1^{v}.

34. **Trinity College 31** Saec. XII^2_4
Isidore, Comm. on the Pentateuch (270 × 175)

Fine arabesque initials. [Illus.]

Prov. Ex Libris of St. Mary Overy, Southwark, fols. 2^{v}. 104. Given by the founder, sir Thomas Pope (d. 1559), fols. 2, 104.

35. **Corpus Christi College 118,** fols. 3–56^{v}
Saec. XII^2_4

Boethius, de musica (210 × 132)

Initials, diagrams. Cf. no. 367.

Prov. Dr. John Dee, 1573, fol. 2^{v}.

James, *Dr. John Dee*, 29.

36. **Christ Church 95** Saec. XII^2_4
St. Paul, Epistles with gloss (252 × 155)

Fine initials (drawings). [Illus.]

Contemporary binding of white leather on boards.

Prov. John Hawson, 1588, fol. i^{v} (bp. of Oxford, 1618).

Boase, *English Art*, 99–100, pl. 33a; G. Pollard, 'The construction of English twelfth-century bindings', *The Library*, XVII (1962), pl. II.

37. **Balliol College 6** Saec. XII^2_4
Augustine, Comm. on St. John's Gospel (385 × 250)

Fine initials (mutilated). Fine marginal drawing, fol. 321, saec. XII^2. [Illus.]

Prov. Note, fol. 226, and chapter numbers in the same hand as in Bodl. MS. Laud Misc. 469, saec. XIV. William Gray, bp. of Ely, 1454–78, fol. i^{v}.

N. R. Ker, *English Manuscripts in the century after the Norman Conquest*, 1960, 7, 53.

38. **Jesus College 52** 1147–76
Cirencester

Bede, de tabernaculo (280 × 195)

Fine arabesque initials. Written under abbot Andrew by Alexander, canon, later chantor, fols. 2–41^{v}, and by Ralph Pulleham, 'scriptor', fols. 42–131^{v}.

Contemporary binding of white leather on boards.

Prov. Ex libris of Cirencester abbey, fol. 1^{v}. Sir John Prise (d. 1555).

Boase, *English Art*, 180, n. 4; N. R. Ker, 'Sir John Prise', *The Library*, X (1955), 13 (J. 20); *id.*, *English Manuscripts in the century after the Norman Conquest*, 1960, 11, 44, 46–7, pl. 20.

39. **Jesus College 67** 1147–76
Cirencester

Bede, Comm. on St. Mark's Gospel (305 × 200)

Fine arabesque initials. Written under abbot Andrew by dominus Odo de Wica, canon.

Contemporary binding of white leather on boards.

Prov. Ex libris of Cirencester abbey, fol. i. Sir John Prise (d. 1555).

N. R. Ker, 'Sir John Prise', *The Library*, x (1955), 13 (J. 11); *id., English Manuscripts in the century after the Norman Conquest*, 1960, 44; J. J. G. Alexander, 'Scribes as artists', *Medieval Scribes, Manuscripts and Libraries. Essays presented to N. R. Ker, eds.* M. B. Parkes, A. G. Watson, 1978, 104, pl. 11d, e.

40. **Jesus College 63** 1147–76
Cirencester

Hegesippus, de excidio Judeorum (325 × 215)

Fine and other initials. Written under abbot Andrew by Ralph Pulleham, 'scriptor', fol. 3. [Illus.]

Medieval binding of white leather on boards, tabs. Repaired.

Prov. Ex libris of Cirencester abbey, fol. 3. Sir John Prise (d. 1555).

T. J. Fordyce, Sir T. M. Knox, 'The Library of Jesus College, Oxford', *Oxford Bibliographical Society Proceedings*, v (1940), 68; Boase, *English Art*, 180 n. 1; N. R. Ker, 'Sir John Prise', *The Library*, x (1955), 13 (J. 13); *id., Medieval Libraries*, 52, 248; *id., English Manuscripts in the century after the Norman Conquest*, 1960, 11, 44–5; J. J. G. Alexander, 'Scribes as artists', *Medieval Scribes, Manuscripts, and Libraries. Essays presented to N. R. Ker, eds.* M. B. Parkes, A. G. Watson, 1978, 104.

41. **Jesus College 70** 1147–76
Cirencester

Bede, Comm. on the Canonical Epistles (280 × 175)

Fine arabesque initials. Written under abbot Andrew by dominus Fulco, canon, fol. i.

Contemporary binding of white leather on boards.

Prov. Ex libris of Cirencester abbey, fol. i. Sir John Prise (d. 1555).

N. R. Ker, 'Sir John Prise', *The Library*, x (1955), 13 (J. 22 ?); *id., English Manuscripts in the century after the Norman Conquest*, 1960, 44; G. Pollard, 'The construction of English twelfth-century bindings', *The Library*, XVII (1962), 16 n. 4, pl. II.

42. **Jesus College 62** 1147–76
Cirencester

Orosius, Historia adversus paganos (325 × 220)

Fine and other initials. Fine arabesque initial. Written under abbot Andrew. [Illus.]

Remains of contemporary white leather binding pasted on more recent cover, tab, original boards.

Prov. Ex libris of Cirencester abbey, fol. 1. Sir John Prise (d. 1555).

Boase, *English Art*, 179 n. 3; N. R. Ker, 'Sir John Prise', *The Library*, x (1955), 13 (J. 25); *id., English Manuscripts in the century after the Norman Conquest*, 1960, 11, 44–5.

43. **Jesus College 53** 1147–76
Cirencester

Bede, Comm. on Samuel (incomplete), etc. (315 × 217)

Good arabesque initials. Written under abbot Andrew by dominus Fulco, canon.

Medieval binding of white leather on boards, much repaired.

Prov. Ex libris of Cirencester abbey, fol. 159v. Sir John Prise (d. 1555).

N. R. Ker, 'Sir John Prise', *The Library*, x (1955), 13, 19 (J. 27).

44. **St. John's College 11** *c.* 1150
Reading

Homiliary (370 × 255)

Fine arabesque initials. Marginal sketches, scribbles, saec. XIII/XIV.

Prov. Reading abbey lists of passages for reading, fol. 1, and of books to be kept in the dormitory, fol. 3, added saec. XIV. Lection for the anniversary of Henry I, founder of Reading abbey, fol. 240. Given by John Stonor of North Stoke, Oxon., 1610, fol. 2.

J. R. Liddell, 'Some notes on the library of Reading Abbey', *B.Q.R.*, VIII (1935), 47, 49, 54.

45. **St. John's College 38** *c.* 1150

Jerome, Comm. on the Psalter (330 × 200)

Arabesque initials.

Prov. Richard Butler, archdeacon of Northampton (d. 1612); given through the good offices of John Buckeridge, bp. of Rochester and Ely, in 1613, fol. 2.

46. **Magdalen College Lat. 31** *c.* 1150

Augustine, de concordia Evangelistarum. Bede, de tabernaculo (262 × 182)

Fine historiated and other initials; good arabesque initials. [Illus.]

Original binding of white leather on oak boards, repaired and resewn.

Prov. Arms of Thomas Tresham, with motto: 'Fecit mihi magna qui potens est, Jun. 29, 1585', fol. 96v. Given by Sir Arthur Throckmorton, 1626, fol. 1.

47. **New College 301** *c.* 1150

Jerome, Comm. on St. Matthew's Gospel (265 × 180)

Fine and other initials (one unfinished).

48. **St. John's College 99** *c.* 1150
Bede, Historia ecclesiastica (250 × 170)

Good initials.

Prov. Ex libris of the Cistercian abbey of Jervaulx, Yorks, fol. i^v, saec. XIII/XIV. Given in 1620 by William Laud, president, archbp. of Canterbury 1633–45, fol. i^v.

Exh., *Archbishop Laud*, no. 8.

49. **Balliol College 218,** ff. 2–70 Saec. XII med.
Julianus Pomerius, Prognosticon, etc. (263 × 165)

Good and other initials, related to St. Albans initial style. [Illus.]

Prov. William Gray, bp. of Ely 1454–78, fol. 1^v.

50. **Trinity College 33** Saec. XII med. Lanthony
Jerome, Comm. on St. Matthew's Gospel (304 × 200)

Good arabesque initials. Written by the same hand as Lambeth Palace MS. 189 and Bodl. MS. Rawl. A. 374, both from Lanthony, Glos.

Prov. Given in 1633 by Francis Baber, fellow, chancellor of the diocese of Gloucester (d. 1669), fol. 3.

N. R. Ker, *English Manuscripts in the century after the Norman Conquest*, 1960, x, pl. 134, Pächt, Alexander, 3, no. 152.

51. **Jesus College 65** Saec. XII med. Gloucester.
Bede, Comm. on the Apocalypse, etc. (288 × 250)

Good arabesque and other initials.

Contemporary binding of white leather on boards, tab.

Prov. 'Ernulfi', first pastedown, saec. XII (cf. Bodl. MS. Laud Misc. 123). Sir John Prise (d. 1555).

N. R. Ker, 'Sir John Prise', *The Library*, X (1955), 13, 19 (J. 19); Ker, *Medieval Libraries*, 92, 265; G. Pollard, 'The construction of English twelfth-century bindings', *The Library*, XVII (1962), 1 n. 2, pl 1.

52. **Trinity College 63** Saec. XII med. Reading
Augustine, Sermons (damaged by damp) (325 × 270)

Fine and other initials close in style to MSS. from Reading, e.g. Bodl. MS. Bodley 257 (cf. Pächt, Alexander, 3, no. 140, pl. XV). [Illus.]

Prov. Reading abbey.

53. **St. John's College 152,** fols. 1^v–56^v Saec. XII med. Canterbury (?)
Priscian, de constructione with gloss (212 × 140)

Initial. Drawing added, fol. 39^v, saec. XIII in. Later marginal scribbles.

Prov. Ex libris of St. Augustine's, Canterbury, pressmark 'D.XI.G.4', fol. i. No. 1390 in the catalogue of saec. XV ex.

James, *Ancient Libraries*, 358; C. R. Dodwell, *The Canterbury School of Illumination 1066–1200*, 1954, 123; M. T. Gibson, 'Priscian, 'Institutiones Grammaticae': a handlist of manuscripts', *Scriptorium*, XXVI (1972), 117.

54. **Balliol College 317** Saec. XII med.
Boethius, de musica (242 × 142)

Fine arabesque initials. Good diagrams.

Prov. Bequeathed by magister Peter of Cossington (d. 1276), fol. 1^v.

Emden, 530–1; Mynors, *Catalogue*, xii; Ker, *College Libraries*, 298 n. 27.

55. **Trinity College 45** Saec. XII med.
Esther, Tobit, Judith, with gloss (300 × 197)

Fine initials. [Illus.]

56. **St. John's College 129** Saec. XII med.
Gospel of St. John with gloss (225 × 120)

Initials. Rough drawing, fol. 84^v, saec. XV^2. Marginal scribbles.

57. **St. John's College 115,** fols. 1–145 Saec. XII med.
Jerome, de viris illustribus (255 × 165)

Good arabesque initials.

Prov. Annotated by Thomas Gascoigne, chancellor of Oxford University (d. 1458), fols. 1, 145.

W. A. Pronger, 'Thomas Gascoigne', *English Historical Review*, LIII (1938), 623 n. 6.

58. **Trinity College 52** Saec. XII med.
Paterius, Liber testimoniorum veteris testamenti, etc. (305 × 200)

Fine arabesque initials (mutilated).

Prov. Given by the founder, sir Thomas Pope (d. 1559), fol. 1.

59. **Magdalen College Lat. 70** Saec. XII med.
Eusebius, Historia romana. Canonical epistles with gloss (220 × 176)

Fine historiated initial. Good and other initials. [Illus.]

60. **Magdalen College Lat. 105** Saec. XII med.
Bede, Historia ecclesiastica (295 × 210)

Fine arabesque initials. [Illus.]

Prov. Possibly from a Cistercian house (evidence of punctuation fols. 1–83). Belonged to Magdalen College by 1600.

Ker, *Catalogue*, no. 357; B. Colgrave, R. A. B. Mynors, *Bede's Ecclesiastical History*, 1969, lvi.

61. **All Souls College 18** Saec. XII med.
Gregory the Great, Epistles (337 × 218)

Fine arabesque initials.

Prov. Given in 1439 by Thomas Gascoigne, chancellor of Oxford University (d. 1458), fols. iv, 1.

W. A. Pronger, 'Thomas Gascoigne', *English Historical Review*, LIII (1938), 618 n. 1; Emden, 746.

62. **Queen's College 309** *c.* 1150–60
Augustine, Sermons (360 × 260)

Good arabesque initial.

Prov. Ex libris of Lanthony, Glos., fol. i, saec. XIII. Lent by the prior of Lanthony to frater Thomas Bradefeld in 1381, fol. i. John Showe, fellow, 1617, fol. i.

63. **Magdalen College Lat. 102** *c.* 1160–70
North England
Ambrose, Augustine, and others, Homilies (303 × 200)

Good and other initials. Marginal drawings, scribbles.

64. **Christ Church 88** 1167
Buildwas

Augustine, Homilies (410 × 295)

Good arabesque initials.

Medieval boards with white leather binding, resewn.

Prov. Ex libris of Buildwas abbey, 1167, fol. 1^v.

Boase, *English Art*, 154 n. 3; N. R. Ker, *English Manuscripts in the century after the Norman Conquest*, 1960, 3, 39, pl. 17.

65. **St. John's College 26, 27** *c.* 1170
St. Albans (?)

Isaiah, Daniel, Ezechiel, with gloss (335 × 212)

Fine historiated and other initials. By the so-called 'Simon painter' who worked at St. Albans under abbot Simon 1167–83. [Illus.]

Prov. Richard Butler, archdeacon of Northampton (d. 1612); given through the good offices of John Buckeridge, bp. of Rochester and Ely, in 1613, fols. 1 (26), 2 (27).

W. Cahn, 'St. Albans and the Channel style in England', *The Year 1200. A Symposium* (The Metropolitan Museum of Art, New York), 1975, 194, 211 no. 7, fig. 27; Thomson, *St. Albans*, 127.

66. **New College 129** Saec. XII$\frac{3}{4}$
Jerome, Epistles (382 × 250)

Fine and other initials. Stub of an initial added in saec. XV, fol. 3. Marginalia may be by the same hand as in Bodl. MS. Auct. E. inf. 1.

Medieval binding of faded pink leather under white leather chemise cut down. Traces of five bosses on each cover. Donor's label under horn cover.

Prov. Given by William Warham, fellow, archbp. of Canterbury, 1503–32.

Emden, 1990; N. R. Ker, *English Manuscripts in the century after the Norman Conquest*, 1960, 41 n. 6, 52.

67. **Corpus Christi College 147** Saec. XII$\frac{3}{4}$
Reading

Cassian, Collationes (275 × 200)

Fine and other initials in Reading style. [Illus.]

Medieval binding of white leather on boards, tabs and strap, English, saec. XIII in.

Prov. Evidence of script and illumination suggests Reading abbey. Possibly in the catalogue of saec. XV in. College ownership 1609, fol. 1.

S. Barfield, 'Lord Fingal's cartulary of Reading Abbey', *English Historical Review*, III (1888), 119.

68. **Balliol College 13** Saec. XII$\frac{3}{4}$
Humberstone (?)

Robert of Tombelaine, Comm. on the Song of Songs (215 × 150)

Fine and other arabesque initials. Good marginal drawings, fols. 18, 24, saec. XII^2.

Prov. Contemporary *ex libris* of Humberstone abbey, Lincs., fol. 1. Probably given by William Gray, bp. of Ely 1454–78.

69. **Lincoln College Lat. 63** Saec. XII$\frac{3}{4}$
Durham

Isidore, Etymologiae (320 × 203)

Fine initial. Fine arabesque initials. Related in style to Durham MSS. [Illus.]

Binding of white leather on original boards (damaged).

Prov. Ex libris of Worksop Priory, Notts. Lent in 1457 by Charles Flemmyng, prior of Worksop, to Robert Flemmyng, dean of Lincoln (d. 1483), who gave it to the College, fol. 5^v.

Ker, *Medieval Libraries*, 215, 320; Weiss, *Humanism*, 103 n. 4.

70. **Corpus Christi College 224,** fols. 142–188^v
Saec. XII$\frac{3}{4}$
York (?)

Boethius, de musica (218 × 138)

Fine and other initials (unfinished). Diagrams. [Illus.]

Medieval binding of white leather on boards.

Prov. Ex libris of St. Mary's, York, fol. 2, saec. XIII. Pressmark of St. Mary's: 'In viij G', fol. 1^{v}, saec. XV. Dr. John Dee, fol. 1.

James, *Dr. John Dee*, 32.

71. **Balliol College 240**, fols. 4–55, 137–89^{v} Saec. XII$\frac{3}{4}$
Religious miscellanea (286 × 200)

Good arabesque and other initials.

Prov. Internal evidence for Monks Kirby priory, Warwicks. Erased *ex libris*, saec. XII/XIII, fol. 137. Richard Bole, fellow, archdeacon of Ely (d. 1477), fol. 3^{v}.

72. **All Souls College 36** Saec. XII$\frac{3}{4}$
Henry of Huntingdon, Historia. Roger Hoveden, Annals, etc. (238 × 160)

Arabesque initials. Marginal drawing, fol. 102.

73. **Balliol College 223** Saec. XII$\frac{3}{4}$
Westminster (?)
Laurence of Westminster and others, Sermons (310 × 220)

Arabesque intials. Marginal drawing, saec. XIII, fol. 16^{v}. Presumably written at Westminster abbey.

Prov. Westminster abbey, London (evidence of contents).

Ker, *Medieval Libraries*, 197 n. 1; R. M. Thomson, *Manuscripts from St. Alban's Abbey 1066–1235*, 1982, 138 n. 30.

74. **Corpus Christi College 189** Saec. XII$\frac{3}{4}$
Constantinus Africanus, Medical treatises (237 × 160)

Good arabesque initials.

Medieval binding of white leather.

Prov. Given to Christ Church, Canterbury, by dominus Johannes Holyngborne, 'monachus', 1510, fol. 1. Dr. John Dee (d. 1608).

James, *Dr. John Dee*, 16.

75. **Corpus Christi College 31** Saec. XII$\frac{3}{4}$
Hilary, de Trinitate (335 × 220)

Fine initial. Good arabesque initial. [Illus.]

Stamped leather binding on boards by the 'Greyhound binder', saec. XV ex.

Prov. Faversham abbey, Kent, first pastedown, saec. XIII^{2}. Given by the founder, Richard Foxe, bp. of Winchester 1501–28, fol. 1.

J. B. Oldham, *English blind-stamped bindings*, 1952, 21 n. 3, pl. XV; Emden, 717.

76. **Merton College 311** (H.3.11) Saec. XII$\frac{3}{4}$
Cicero, de officiis, etc. (250 × 172)

Good arabesque initials.

Prov. List of members of an unidentified ecclesiastical household, saec. XIV in, fol. 100^{v}. Bought from Thomas Trillek, bp. of Rochester (d. 1372) by William Reed, bp. of Chichester 1368–85, who gave it to the College.

Exh., *The Survival of Ancient Literature*, Bodleian Library, Oxford, 1975, no. 130, Appendix, 106.

77. **Magdalen College Lat. 119** Saec. XII$\frac{3}{4}$
Psalter (part) with comm. by Gilbert de la Porrée (295 × 210)

Good arabesque initials.

Prov. 'Memoriale magistri Johannis de Hamt . . .', fol. 1, saec. XIII.

Ker, *Medieval Libraries*, 329.

78. **Magdalen College Lat. 84** Saec. XII$\frac{3}{4}$
Ivo of Chartres, Chronica Franciae (282 × 192)

Fine initial. Arabesque initials. [Illus.]

Original boards with a later tanned leather cover.

79. **Merton College 316** (M.2.2) Saec. XII$\frac{3}{4}$
Josephus, de bello Judaico (320 × 200)

Fine arabesque initials.

Prov. Given in 1493 by the executors of John Tailour, provost of Oriel College, chancellor of Exeter (d. 1492), fol. 1^{v}.

Emden, 1851; N. R. Ker, 'Oxford College libraries in the sixteenth century', *B.L.R.*, VI (1957–61), 473.

80. **Merton College 88** (B.1.11) Saec. XII$\frac{3}{4}$
English (?)
Hugh of Fleury, Chronicon. Bede, Chronica (279 × 183)

Good arabesque initials.

Prov. Bequeathed by John Reynham, fellow, chancellor of Oxford University 1361–3 (d. 1376), fol. 1.

Emden, 1570.

81. **Queen's College 306** Saec. XII$\frac{3}{4}$
Jerome, Comm. on Jeremiah (incomplete) (390 × 278)

Good and other arabesque initials.

Prov. Johannes Shaw, 1622, fol. 1.

82. **St. John's College 95** Saec. XII$\frac{3}{4}$
Orosius, Historia adversus paganos (265 × 170)

Fine initial. Fine arabesque initials. [Illus.]

Prov. Pressmark 'I.xvii' of Chichester cathedral, fol. 1, saec. XIV.

83. **New College 32** Saec. XII$^{3}_{4}$
Peter Lombard, Comm. on the Psalter (320 × 220)

Fine and other initials.

Prov. Given by William Ive, headmaster of Winchester College (d. 1486), fol. iii^{v}.

Emden, 1009.

84. **St. John's College 96** Saec. XII$^{3}_{4}$
Lives of Saints (250 × 170)

Good initial, good and other arabesque initials. Close in style to the Winchcomb Psalter, Dublin, Trinity College MS. 53 (cf. also Bodl. MS. Douce 368). [Illus.]

Prov. Given to Pershore abbey by Gillebertus, prior, saec. XIII (?), fol. i. Dominus Symon Evyrton, choral vicar of Wells cathedral, saec. XIV, fol. ii.

N. R. Ker, *English Manuscripts in the century after the Norman Conquest*, 1960, 20 n. 5.

85. **University College 115** Saec. XII$^{3}_{4}$
Bible (Genesis-Jeremiah) (338 × 260)

Fine and other initials (unfinished).

Hunt, *Manuscripts*, 19 n. 2, 27.

86. **Trinity College 34** Saec. XII$^{3}_{4}$
English (?)

Berengaudus, Comm. on the Apocalypse. Misc. religious treatises (305 × 202)

Fine initial. Good arabesque initials. Some flourishes added later.

Prov. Ex libris of the Cistercian abbey of Kingswood, Glos., saec. XII/XIII, fol. 2^{v}.

87. **Trinity College 69** Saec. XII$^{3}_{4}$
Jerome, Comm. on Minor Prophets (335 × 225)

One good and other initials. Initial, fol. 19^{v}, with later additions, saec. XIII (?).

Prov. Contemporary *ex libris* of Lanthony priory, Glos., fol. ii. Given in 1633 by Francis Baber, fellow, chancellor of the diocese of Gloucester (d. 1669), fol. 1.

88. **Trinity College 58** Saec. XII$^{3}_{4}$
Psalter with comm. by Gilbert de la Porrée (incomplete) (310 × 215)

Fine historiated initials (mutilated), fol. 180 very similar to an initial in Bodl. MS. Auct. D.2.1, fol. 147^{v}. Good arabesque initials. [Illus.]

Prov. Given by fr. Thomas Walden (Thomas Netter of Walden), prior provincial of the Carmelites in England (d. 1430), to the Carmelite Convent in London, fol. i^{v}. Pressmark 'C.24^{m}', fol. 1. Given by the founder, sir Thomas Pope (d. 1559), fols. iv, 1, 251^{v}.

Emden, 1344; Ker, *Medieval Libraries*, 125, 279; Pächt, Alexander, 3, no. 226; Morgan, *Early Gothic*, 54.

89. **St. John's College 49** Saec. XII$^{3}_{4}$
Peter Lombard, Sentences (305 × 250)

Fine historiated and other initials. Initial, fol. 3^{v}, replacement of saec. XV. [Illus.]

Prov. Hilary, bp. of Chichester (d. 1169), fol. 2. Chichester cathedral pressmark 'F xxiiii', saec. XV^{2}, fol. 2. Given in 1617 by William Laud, president, archbp. of Canterbury 1633–41, fol. 1^{v}.

Exh., *Archbishop Laud*, no. 4; M. B. Parkes, 'The influence of the concepts of *Ordinatio* and *Compilatio* on the development of the Book', *Medieval Learning and Literature. Essays presented to R. W. Hunt, eds.* J. J. G. Alexander, M. T. Gibson, 1976, 139, pl. x.

90. **St. John's College 111** Saec. XII$^{3}_{4}$
Gospel of St. Matthew with gloss of Anselm of Laon (243 × 150)

Drawings (unfinished) added, fols. 19^{v}, 107^{v}, to a French MS. of saec. XII$^{2}_{4}$. [Illus.]

Prov. Ex libris of Isle of May priory, Fife, saec. XII–XIII, fol. iii^{v}. Given by William Laud, president, archbp. of Canterbury 1633–41, fols. 1, 2.

Exh., *Archbishop Laud*, no. 3.

91. **New College 363**, fols. 4, 5. Saec. XII$^{3}_{4}$
All Souls College 401, fols. i, 109.

Hugh of St. Victor, de sacramentis (fragments) (approx. 305 × 235)

Fine and other arabesque initials. [Illus.]

Prov. 'Iste liber pertinet ad me Johannem . . . et ad fratrem . . .', New College, fol. 4. All Souls College leaves used in the binding of the College building accounts, 1593–4. Other parts of this MS. are in Merton College E.3.30 and London, Lambeth Palace 41. C.3.

Ker, *Pastedowns*, nos. 932, 957, 1010, etc.

92. **Corpus Christi College 16** Saec. XII$^{3}_{4}$
Minor Prophets with gloss (272 × 195)

Good arabesque initials. Pen flourishes added on fol. 2^{v}, saec. XIII.

Stamped leather binding on boards by the 'Greyhound binder', clasps, saec. XV ex., Oxford (?).

Prov. Given by the founder, Richard Foxe, bp. of Winchester 1501–28, fol. 2.

J. B. Oldham, *English blind-stamped bindings*, 1952, 21 n. 3, pl. XV; Emden, 717.

93. **All Souls College 30** *c.* 1170–80
Bernard of Clairvaux, Sermons on the Song of Songs (410 × 280)

Good arabesque initials.

Medieval binding of white leather on boards, originally with five bosses on each cover.

Prov. Marginalia possibly by the founder, Henry Chichele, archbp. of Canterbury 1414–43. Given by Richard Andrew, warden 1437–42, dean of York (d. 1477).

94. **Oriel College 36,** fols. 143–323^{v} *c.* 1170–80
Augustine, Comm. on Pss. LI–LXXX (415 × 285)

Good initials. Fragment inserted in a MS. of saec. XIV^{1} with fine penwork initials.

Cover of old binding preserved as fly-leaves.

Prov. Thomas Gascoigne, chancellor of Oxford University (d. 1458), fol. 143.

W. A. Pronger, 'Thomas Gascoigne', *English Historical Review*, LIII (1938), 618 n. 5; Emden, 746.

95. **Merton College 177** (A.1.5) *c.* 1170–80
North England

Bede, Homilies (295 × 205)

Fine arabesque initials related in style to Durham MSS.

Binding of blind-stamped leather on boards, *c.* 1500.

Prov. Richard Hakeborne, fellow (d. 1322), last pastedown. Bequeathed *c.* 1349 by John of Staveley, fellow, fol. 2^{v}.

96. **New College 35** *c.* 1170–80
North England

Peter Lombard, Comm. on the Psalter (350 × 235)

Fine and good initials. Related in style to Durham MSS.

Prov. Thomas Hylle, fellow, rector of St. Ebbe's, Oxford (d. 1469). Given by his executors to Henry Upnore, fellow, who bequeathed it to the College (d. 1488), fol. 1^{v}.

Emden, 935, 1932.

97. **Balliol College 188** *c.* 1175–85
Ambrosius Autpertus, Comm. on the Apocalypse, etc. (incomplete) (328 × 222)

Fine arabesque initials.

Prov. 'Liber sce Marie de . . .', fol. 2^{v}, partly erased *ex libris*, probably of Fountains abbey (cf. Oxford, Bodl. MS. Laud Misc. 310).

98. **Balliol College 9** *c.* 1180
Augustine, Comm. on Genesis (290 × 190)

Arabesque initials. Marginal drawing, fol. 43.

Prov. William Gray, bp. of Ely 1454–78, fol. 1.

99. **Balliol College 156** *c.* 1180
Jerome, Comm. on Isaiah (383 × 286)

Fine arabesque initials. [Illus.]

Prov. William Gray, bp. of Ely 1454–78, fol. 1^{v}. Annotations by Thomas Gascoigne, chancellor of Oxford University (d. 1458), *passim.*

100. **Corpus Christi College 209** Saec. XII ex.
Fountains

Augustine, Soliloquia. Consuetudines Cisterciensium (226 × 155)

Good arabesque initials.

Contemporary binding of white leather on boards with seal skin chemise, metal bosses on each cover.

Prov. Contemporary *ex libris* of Fountains abbey, Yorks., fol. 2. Pressmark 'Th. 3 in ant lib. 38', first pastedown, saec. XV in. Given by John Rosewell, headmaster of Eton College (d. 1684), fol. 1.

G. Pollard, 'The construction of English twelfth-century bindings', *The Library*, XVII (1962), 1, 13, pls. I, II; *id.*, 'Describing medieval bookbindings', *Medieval Learning and Literature. Essays presented to R. W. Hunt, eds.* J. J. G. Alexander, M. T. Gibson, 1976, 58.

101. **St. John's College 43** Saec. XII ex.
Bury St. Edmunds

Peter Lombard, Comm. on St. Paul's Epistles (310 × 235)

Fine historiated (one) and other initials. Arabesque initial. [Illus.]

Prov. Ex libris of Bury St. Edmunds abbey, letter mark 'P. 98', fol. 1, saec. XIII. Pledge of fr. J. Haliwelle, monk of Bury St. Edmunds, fol. 234^{v}, saec. XIV (?).

102. **Jesus College 105** Saec. XII ex.
Hereford (?)

Gospel of St. Luke with gloss (303 × 204)

Fine arabesque initials.

Medieval binding of white leather on boards, chemise sewn down.

Prov. Given to St. Guthlac's priory, Hereford, by Roger, 'vicedecanus', saec. XIII. Pressmark 'XXus VIus', fol. 2.

N. R. Ker, 'Medieval pressmarks of St. Guthlac's Priory (Hereford) and Roche Abbey, Yorks.', *Medium Aevum*, V (1936), 47–8.

103. **Magdalen College Lat. 128** Saec. XII ex.
Osney

Peter Lombard, Comm. on the Psalter (332 × 230)

Good arabesque initials (mutilated).

Prov. Pledged in the Rothbury chest, Oxford University, 1458, fol. 270^v. John Grene, headmaster of Winchester College (d. by 1484), fol. i^v.

Emden, 816; Ker, *Medieval Libraries*, 141 n. 8.

104. **Balliol College 173B** Saec. XII ex. Buildwas (?)
Peter Lombard, Comm. on St. Paul's Epistles (370 × 265)

Fine initials (mutilated) with woven textile covers. [Illus.]

Prov. Ex libris of Buildwas abbey, Shropshire. (since removed, cf. Mynors, *Cat.*). William Gray, bp. of Ely, 1454–78, fol. 206.

105. **Balliol College 35A** Saec. XII ex.
Peter Lombard, Comm. on the Psalter (394 × 295)

Fine historiated and other initials (mutilated). Close in style to the main hand of the 'Laud Bible' (Pächt, Alexander, 3, no. 254). Fine arabesque initials. [Illus.]

Prov. Bequeathed by magister Walterus de Brug', 'dictus le Paumar', to Buildwas abbey, Shropshire, under abbot Willelmus Tyrry, 1277, fol. 3. Bequeathed by Robert Thwaites, master, dean of Auckland, co. Durham (d. 1458), fols. 1^v, 4.

Mynors, *Catalogue*, xx–xxi.

106. **New College 308** Saec. XII ex.
Bede, Historia ecclesiastica (350 × 265)

Fine historiated initials. Fine penwork initials. List of popes from Peter to Innocent III (d. 1216) added on fols. 2^v–3^v, saec. XIII. [Illus.]

Prov. The Cistercian pope, Eugenius III (1145–53), marked 'nostri ordinis' in the original hand, suggests a Cistercian house.

B. Colgrave, R. A. B. Mynors, *Bede's Ecclesiastical History*, 1969, lii.

107. **Magdalen College Lat. 17** Saec. XII ex.
Priscian, Institutiones grammaticae (250 × 175)

Fine and other arabesque initials in two styles (mutilated). [Illus.]

Prov. '. . . de sorte Collett' (prob. William Collet, fellow, d. by Apr., 1500), fol. i.

Emden, 464; N. R. Ker, *Medieval Manuscripts in British Libraries*, I, 1969, 402–3; M. T. Gibson, 'Priscian, 'Institutiones Grammaticae': a hand list of manuscripts', Scriptorium, XXVI (1972), 117.

108. **Lincoln College Lat. 13** Saec. XII ex.
Augustine, Comm. on Pss. CI–CL (340 × 230)

Arabesque initials. Initial, fol. 1, added saec. XIII^2.

109. **Balliol College 257** Saec. XII ex.
Euclid, Elements (220 × 157)

Good initials. Marginal diagrams.

110. **Exeter College 25** Saec. XII ex.
Josephus, Antiquitates Judaicae (incomplete) (340 × 225)

Arabesque and penwork initials.

Prov. Given by dominus John French, chaplain of Eton College (d. 1457), fol. 130.

Emden, 726.

111. **Lincoln College Lat. 30** Saec. XII ex.
Bede, Homilies (200 × 102)

Initial. Penwork initials.

Prov. Given by the founder, Richard Flemmyng, bp. of Lincoln (d. 1431), fol. 1^v.

Emden, 698.

112. **Lincoln College Lat. 17** Saec. XII ex.
Haimo, Comm. on Isaiah (280 × 200)

Arabesque and penwork initials.

Prov. Notes by Thomas Gascoigne, chancellor of Oxford University (d. 1458), fol. 1.

113. **Magdalen College Lat. 118** Saec. XII ex.
Gilbert de la Porrée, Comm. on St. Paul's Epistles (330 × 210)

Fine arabesque initials. [Illus.]

Original boards with later blind-stamped leather cover.

Prov. John Budden, D.C.L. (d. 1620), fol. 1.

114. **Balliol College 176** Saec. XII ex.
Bede, Historia ecclesiastica, etc. (360 × 250)

Good arabesque initials.

Prov. William Gray, bp. of Ely 1454–78, fol. 1^v.

115. **Balliol College 23** Saec. XII ex.
Peter Cantor, Comm. on the Bible (Numbers-Ruth) (375 × 263)

Arabesque initials.

Prov. Pledged by W. Collyng, fellow, in the Burnell chest, Oxford, for xix s., iiij d., 1432, fol. 148^v. William Gray, bp. of Ely 1454–78, fol. 1^v.

Mynors, *Catalogue*, xli.

116. **All Souls College 12** Saec. XII ex.
Anonymous, Comm. on the Pentateuch (287 × 190)

Good initial (drawing). Good arabesque initials.

Medieval binding of white leather on boards (rebacked).

Prov. Ex libris of Stanley abbey, Wilts. (or Stoneleigh, Warwicks.), fol 180^{v}.

Ker, *Medieval Libraries*, 183 n. 1.

117. **New College 34** Saec. XII ex.
Peter Lombard, Comm. on the Psalter (415 × 268)

Fine and other initials.

Prov. Henry Beaufort, bp. of Winchester 1404, nom. cardinal 1426 (d. 1447), fol. ii^{v}. Given by Thomas Chaundler, fellow, chancellor of Oxford University 1457 (d. 1490).

M. R. James, *The Chaundler MSS.* (Roxburghe Club), 1916, 5; Emden, 142, 399; Weiss, *Humanism*, 133 n. 8.

118. **Balliol College 178B** Saec. XII ex.
Florus diaconus, Comm. on St. Paul.s Epistles. Part II. (incomplete) (440 × 304)

Fine initials. Companion volume to MS. 178A (no. 33). [Illus.]

Prov. William Gray, bp. of Ely 1454–78, fol. 1^{v}.

119. **Balliol College 10** Saec. XII ex.
Augustine, Comm. on the Trinity (340 × 220)

Good arabesque initials.

Prov. William Gray, bp. of Ely 1454–78, fol. 1^{v}.

120. **Corpus Christi College 15** Saec. XII ex.
Genesis, with gloss (320 × 222)

Fine initial. One good arabesque initial with later additions.

121. **New College 104** Saec. XII ex.
Peter Comestor, Historia scholastica, part II (412 × 292)

Good and other initials. Good marginal drawing, fol. 37. Tables of consanguinity, diagram, fols. 118–21, added saec. XIII$\frac{2}{4}$.

Prov. Given by Thomas Cranley, warden, archbp. of Dublin 1397–1417, who bought it from an Oxford stationer, John Brown.

Emden, 511.

122. **Jesus College 108** Saec. XII ex.
Acts of the Apostles, with gloss (235 × 158)

Good initial.

123. **Corpus Christi College 52** Saec. XII ex.
Peter Lombard, Sentences (340 × 224)

Fine historiated (one) and other initials. [Illus.]

Binding of brown leather, original boards, rebound.

Prov. Henry Parry, fellow, 1618, fol. 1.

E. Rathbone, 'Peter of Corbeil in an English setting', *Medieval learning and literature. Essays ppresented to R. W. Hunt, eds.* J. J. G. Alexander, M. T. Gibson, 1976, 288–91, 296–7, 303 ff.

124. **New College 130** Saec. XII ex.
Cyprian, Chrysostom, etc. (442 × 300)

Fine arabesque and penwork initials. Marginal drawings probably added saec. XIII in.

Prov. Probably comes from a Cistercian house. Given by William Say, fellow, dean of St. Paul's, London (d. 1468), fol. 1.

N. R. Ker, 'Thomas James's Collation of Gregory, Cyprian, and Ambrose', *B.L.R.* IV (1952), 27; *Rural Life*, fig. 1b; Emden, 1650; Hunt, *Medieval Library*, 328.

125. **Lincoln College Lat. 27,** ff. 62–84^{v}. Saec. XII^{2}
Statius, Achilleis (220 × 125)

Initials. Marginal drawing. Diagrams. Cf. no. 10.

Medieval binding of white leather on boards.

Prov. Ex libris (obliterated) of Sempringham priory, Lincs., fol. 6^{v}. Robert Flemmyng, dean of Lincoln (d. 1483), fol. 7.

126. **Jesus College 66** Saec. XII^{2}
Hereford (?)
Gospel of St. Matthew with gloss (265 × 180)

Good initial. Good arabesque initials.

Contemporary binding of white leather on boards.

Prov. Given to St. Guthlac's priory, Hereford, by Roger, 'vicedecanus', saec. XIII, fol. 1^{v}. Pressmark 'XLus Vus', fol. 4. Sir John Prise (d. 1555).

N. R. Ker, 'Medieval pressmarks of St. Guthlac's Priory (Hereford) and Roche Abbey, Yorks.', *Medium Aevum*, V (1936), 47–8; *id.*, 'Sir John Prise', *The Library*, V (1955), 13 (J. 9).

127. **Magdalen College Lat. 139** Saec. XII/XIII
Alexander Nequam, de naturis rerum (390 × 280)

Fine and other initials. Penwork initials.

Original binding of white leather on boards.

128. **Balliol College 229** *c.* 1200
Buildwas

Jerome, Letters, etc. (332 × 228)

Good arabesque initials.

Prov. Contemporary *ex libris* of Buildwas abbey, Shropshire, fol. 1^{v}. Pledged by M. Johannes . . . in the Langton chest, Oxford University, 1421, fol. i. William Gray, bp. of Ely 1454–78, fol. 1. Notes by Thomas Gascoigne, chancellor of Oxford University (d. 1458).

Weiss, *Humanism*, 91 n. 5.

129. **St. John's College 31** *c.* 1200
Peter of Cornwall, Pantheologus (320 × 230)

Good arabesque initials.

Prov. Ex libris of the Augustinian house of St. Thomas Martyr at Lessnes, Kent, saec. XIII, fol. 1. Richard Tileslye, archdeacon of Rochester, 1619, fol. 2.

130. **New College 101** *c.* 1200
Peter Comestor, Historia scholastica with gloss (325 × 223)

Fine and other initials.

Prov. Deposited by John and Stephen Hormesby as caution for a loan in the Old Oxford University chest, 9 March, 1355, fol. 1^{v}. Caution of 'magistri . . . in cista de Turville, 1356', fol. 1^{v}. Bought by Lewis Charlton, bp. of Hereford 1361–9, from Richard Lynne, a University stationer, 8 Febr., 1358, fol. 2.

Emden, 392, 964.

131. **Corpus Christi College 95,** fols. 59–117 *c.* 1200
Albumazar, Astrologia (225 × 185)

Fine initials, diagrams. Cf. no. 236.

132. **Queen's College 300** *c.* 1200
Kings I–IV, with gloss (incomplete) (327 × 220)

Good initials (mutilated).

133. **New College 210** *c.* 1200
Gratian, Decretum, with gloss of Bartholomew of Brescia (315 × 250)

Fine initials. Penwork initials.

Prov. R. Wylleys, fellow, vicar of Higham Ferrers 1504–23, fol. 2^{v}.

Emden, 2117.

134. **St. John's College 5** *c.* 1200
Rhabanus Maurus, Etymologiae (410 × 285)

Fine border, historiated and other initials. Good penwork initials. [Illus.]

Prov. Ex libris of Reading abbey, fol. i^{v}, saec. XIII. Clemens Burdet, fellow of Queen's College (?), rector of Englefield, Berks., 1542–60/1, fol. 182. Given by John Stonor of North Stoke, Oxon., 1609, fols. i^{v}, 1, 181.

J. R. Liddell, 'Sone notes on the library of Reading Abbey', *B.Q.R.*, VIII (1935), 54.

135. **Trinity College 72** *c.* 1200
Radulphus Flaviacensis, Comm. on Leviticus (315 × 205)

Good and other initials.

Prov. Given by the founder, sir Thomas Pope (d. 1559), fol. 1.

136. **Exeter College 40** *c.* 1200
Gregory the Great, Homilies on Ezechiel (335 × 243)

Good initials.

137. **Lincoln College Lat. 83** *c.* 1200
Peter of Cornwall, Pantheologus (part IV) (325 × 234)

Good arabesque and other initials. [Illus.]

Prov. Magister Thomas Goodard (?), fellow, pasted on fol. 1, saec. XIV (?).

R. W. Hunt, 'English learning in the late 12th century', *Transactions of the Royal Historical Society*, XIX (1936), 33 n. 1.

138. **Merton College 183** (O.3.6) *c.* 1200
Peter Lombard, Comm. on St. Paul's Epistles (347 × 250)

Fine initials. [Illus.]

Medieval binding of white leather on boards, chemise.

139. **Merton College 15** (B.2.3) *c.* 1200
Homilies of Gregory, Augustine and others (410 × 275)

Fine and other initials.

Prov. Given *c.* 1394 by Walter Rammesbury, fellow, precentor of Hereford cathedral (d. 1406), fol. i^{v}.

Emden, 1545.

140. **University College 130** *c.* 1200
English (?)
Bible (Genesis–Ezechiel) (400 × 275)

Fine historiated and other initials. Fine arabesque initials. Marginal sketches, fols. 1^{v}, 124^{v}. [Illus.]

Contemporary binding of white leather on boards.

Prov. Ex libris of St. Neot's priory, Hunts., fol. 171^{v}, saec. XIV.

141. **St. John's College 165** *c.* 1200
Anselm of Canterbury, Prayers and meditations (208 × 138)

Arabesque initials.

Prov. Given to St. Mark's, Bristol, by Savericus de Lya, presbyter, with reversion if claimed, to the executors of Robert de Thornbire, fol. ii, saec. XIII^{2}.

142. **Merton College 9** (A.2.9) *c.* 1200
Origen, Misc. works (325 × 220)

Fine initial. Penwork initials.

Prov. Given by Richard FitzJames, warden 1483–1507, bp. of London (d. 1522), fol. ii^{v}.

Emden, 691–2.

143. **St. John's College 112**, fols. 115–210 Saec. XIII in.

Itinerarium Clementis (242 × 158)

Good arabesque initials. Cf. no. 725.

Prov. Given to Oriel College by William Dalton (d. 1371), fol. 115.

C. L. Shadwell, 'A Catalogue of Oriel College Library in . . . 1375 A.D.', *Collectanea I, ed.* C. R. Fletcher, *Oxford Historical Society*, V (1885), 69.

144. **St. John's College 51** Saec. XIII in.
Alexander Nequam, de naturis rerum, etc. (310 × 205)

Good arabesque and penwork initials. 'Henricus istum scripsit librum', fol. 170v.

Prov. Henry Savile, fol. 1 (d. 1617). Given in 1634 by sir William Paddy, commoner, physician to James I, fol. 2.

A. G. Watson, *The Manuscripts of Henry Savile of Banke*, 1969, no. 9.

145. **Trinity College 74** Saec. XIII in.
Bible (275 × 218)

Fine historiated and other initials. Good penwork initials. [Illus.]

Prov. Pressmark of Bury St. Edmunds 'B.15'. Hymn in honour of St. Edmund, king and martyr, saec. XIII (?), fol. 210. Given by the founder, sir Thomas Pope (d. 1559), fols. 1, 208v.

146. **St. John's College 61** Saec. XIII in.
Bestiary (293 × 198)

Important miniatures, initials. [Illus.]

Prov. Ex libris of Holy Trinity priory, York, saec. XIII, fol. 103v. Given in 1634 by sir William Paddy, commoner, physician to James I, fol. 5.

G. C. Druce, 'The Elephant in Medieval Legend and Art', *Archeological Journal*, XXVI (1919), 35, 44, pl. V no. 2; *id.*, 'The medieval bestiaries and their influence on ecclesiastical decorative art', *Journal of the British Archaeological Association*, XXVI (1920), 35, 60, pls. II, VII; M. R. James, *The Bestiary* (Roxburghe Club), 1928, 18, pls. 17–18; J. Baltrusaitis, *Reveils et Prodiges, le gothique fantastique*, 1960, 130–1, fig. 19c; Morgan, *Early Gothic*, no. 42; 12, 60, 64, 67, 90, fig. p. 8.

147. **St. John's College 21** Saec. XIII in.
Joshua and Judges with gloss (355 × 240)

Fine initials. [Illus.]

Prov. Ex libris of Reading abbey, fol. i, saec. XIII. Given by John Stonor of North Stoke, Oxon., 1609, fol. 1.

148. **New College 274** Saec. XIII in.
St. Albans (?)
Pliny the Elder, Natural History (398 × 260)

Fine and other initials. Good penwork initials. *Mappa mundi*, saec. XIV, fol. IIIv. [Illus.]

Prov. Ex libris of St. Albans abbey, saec. XIII in. (erased), fol. 1. Returned later to the abbey by Richard de Bury, bp. of Durham (d. 1345), fol. iiiv. Pledged between 1453–62 in the Queen's chest, Oxford University, jointly by John Warder, monk of St. Albans, and John Consobrine, a Carmelite, fol. 1. Pledged in the Chichele chest, Oxford, by Robert Woodstock, fol. 97v. Bought in London in 1471 by John Russel, later bp. of Lincoln, who gave it to the College in 1482, fol. iv.

'Richard de Bury Books from the Abbey of St. Albans', *B.L.R.*, III (1950–1), 177–9; Exh., *Oxford College Libraries in 1556*, Bodleian Library, Oxford, 1956, no. 56; Emden, 478, 610, 983; Ker, *Medieval Libraries*, 168, 302; Exh., *The Survival of Ancient Literature*, Bodleian Library, Oxford, 1975, no. 132, pl. XXII; Thomson, *St. Albans*, 73, 75, 112, 136 n. 76, pls. 245-6, 254.

149. **New College 159** Saec. XIII in.
Rhabanus Maurus, de natura rerum (398 × 215)

Fine initial. Fine and other penwork initials, marginal ornament (quire marks).

150. **Exeter College 38** Saec. XIII in.
Augustine, Comm. on Pss. I–L (incomplete) (335 × 223)

Good penwork initials. Marginal drawings by the scribe.

151. **Oriel College 60** Saec. XIII in.
Gregory the Great, Registrum epistolarum (334 × 225)

Fine arabesque and penwork initials (mutilated).

Medieval cover and pastedown preserved as fly-leaves.

Prov. Prefaced by Bernard of Clairvaux, Sermons, fols. iii–vv, saec. XII ex. Lettermark 'X" of the Cluniac priory of St. Andrew, Northampton, saec. XIII, fol. iii.

152. **Pembroke College 3** Saec. XIII in.
Bede, Historia ecclesiastica (incomplete) (350 × 270)

Good arabesque initials. Good penwork initials. Marginal drawing, fol. 83, saec. XV (?).

Medieval boards with white leather cover, saec. XV. Rebound.

Prov. Anthony Cole of Tadwick, fol. 46v, saec. XVI.

B. Colgrave, R. A. B. Bynors, *Bede's Ecclesiastical History*, 1969, lii.

153. **Queen's College 317** Saec. XIII in.
Gospels of Sts. Matthew, Mark, with gloss (293 × 203)

Good historiated initial. Good penwork initials.

Medieval boards under later velvet cover, metal bosses.

Prov. *Ex libris* of St. Mary's abbey, Reading, saec. XIII, fol. i. Henry VIII (evidence of binding).

J. R. Liddell, 'Some notes on the library of Reading Abbey', *B.Q.R.*, VIII (1935), 51.

154. **Queen's College 303** Saec. XIII in.
Peter Lombard, Comm. on the Psalter (355 × 242)

Fine initials (mutilated). Fine penwork initials. [Illus.]

Binding of medieval boards with later velvet cover.

155. **Queen's College 55** Saec. XIII in.
Bible (290 × 192)

Fine penwork initials (mutilated).

Prov. Bequeathed by Thomas Barlow, provost, bp. of Lincoln (d. 1691), to his chaplain, William Offley, who gave it to the College, 1723.

156. **Queen's College 322** Saec. XIII in.
Peter of Poitiers, Distinctiones super Psalterium (250 × 170)

Fine and other penwork initials. Marginal drawing, fol. 136.

Contemporary binding of oak boards.

Prov. 'prioris Willelmi', fol. i^{v}, saec. XIII/XIV.

157. **New College 107** Saec. XIII in.
Peter Lombard, Sentences (330 × 210)

Good initial. Penwork borders, initials. Tables of consanguinity, diagram, fols. 123^{v}–4.

158. **Queen's College 52** Saec. XIII in.
Bible (422 × 200)

Fine historiated (one drawing) and other initials. Fine penwork initials. [Illus.]

Prov. Given by Francis Newport to Ben. Jenks, rector of Harley, co. Salop., who gave it to the College, 1697. fol. i.

H. H. Glunz, *History of the Vulgate in England*, 1933, 317, 389.

159. **Merton College 29** (A.3.3.), fols. 15–173
Saec. XIII in.

Origen, Misc. works (335 × 223)

Fine initials. Good penwork initials.

Prov. Given by John Reynham, fellow, chancellor of Oxford University 1361–3 (d. 1376), fol. 15.

Emden, 1570.

160. **Merton College 182** (Q.1.8) Saec. XIII in.
Peter Lombard, Comm. on St. Paul's Epistles (340 × 243)

Fine historiated and other initials. Fine penwork initials. [Illus.]

Prov. Given by John Burbach, fellow, rector of St. John the Evangelist's, London (d. 1451), fol. ii^{v}.

Emden, 306.

161. **Merton College 208** (N.2.5) Saec. XIII in.
Psalter with gloss (392 × 253)

Fine historiated and other initials (mutilated). Good penwork initials. [Illus.]

Prov. Given by John Burbach, fellow, rector of St. John the Evangelist's, London (d. 1451), fol. 1.

Emden, 306.

162. **Queen's College 316** Saec. XIII in.
Deuteronomy with gloss (290 × 200)

Fine initial.

Contemporary binding of white leather on boards, strap.

Prov. Fr. William Donyngton, 1471, first pastedown.

163. **Magdalen College Lat. 106** Saec. XIII in.
Gregory the Great, Homilies, etc. (340 × 232)

Fine arabesque initials. Good penwork initials. [Illus.]

Contemporary binding of white leather on boards, clasps.

Prov. Given by John Sydnere, fellow 1475.

Emden, 1839.

164. **Merton College 249** (B.1.7), fols. 1.10^{v}
Saec. XIII in.
English (?)

Philip de Thaon, Bestiary. In French (273 × 180)

Good miniatures (drawings, mutilated). [Illus.]

Prov. Given by William Reed, bp. of Chichester 1368–85, who bought it from Thomas Trillek, bp. of Rochester (d. 1372), fols. i, ii^{v}.

Emden, 1559; X. Muratova, 'The decorated manuscripts of the bestiary of Philippe de Thaon', *Proceedings Third International Beast Epic, Fable and Fabliau Colloquium*, Münster 1979, *eds*. J. Goossens, T. Sodmann, 217–46, figs.

165. **Magdalen College Lat. 24** Saec. XIII in.
Petrus de Riga, Aurora (230 × 140)

Fine and other initials. [Illus.]

Prov. Bequeathed by Richard Laugharne, fellow, who bought it from dominus William Demet in 1485, fol. 1.

Emden, 1107.

166. **Corpus Christi College 114** Saec. XIII in.
Aristotle, Corpus vetustius with gloss (306 × 215)

Good initials. [Illus.]

Prov. Henry Parry, fellow, 1619, fol. 5.

Aristoteles Latinus, I, no. 354.

167. **Magdalen College Lat. 124** Saec. XIII in. Osney

Joshua-Kings, with gloss (357 × 240)

Good penwork initials.

Prov. Osney abbey pressmark (D.31), saec. XIII, fol. v^{v}. John Grene, headmaster of Winchester College (d. by 1484).

Emden, 816; Ker, *Medieval Libraries*, 141 n. 8; S. Patterson, 'Comparison of minor initial decoration', *The Library*, XXVII (1972), 23–30.

168. **Magdalen College Lat. 130** Saec. XIII in. Osney

Gospels with gloss (335 × 225)

Fine historiated initials. Fine penwork initials. Canon Tables, fols. v–ix (unfinished). [Illus.]

Prov. Osney abbey pressmark 'R', fol. 1. Given by 'Willelmus de Sancto Martino (?), capellanus', saec. XIII, fol. ii. John Grene, headmaster of Winchester College (d. by 1484), fol. i.

Emden, 816; Ker, *Medieval Libraries*, 148 n. 8, 287.

169. **Magdalen College Lat. 131** Saec. XIII in. Osney

Peter Lombard, Comm. on St. Paul's Epistles (355 × 235)

Good initials (damaged and mutilated).

Prov. Osney abbey pressmark 'X', fol. 1. John Grene, headmaster of Winchester College (d. by 1484).

Emden, 816; Ker, *Medieval Libraries*, 141 n. 8.

170. **Magdalen College Lat. 132** Saec. XIII in. Osney

Acts, Epistles, Apocalypse, with gloss (305 × 200)

Good and other initials.

Prov. Osney abbey pressmark 'T', fol. 1. John Grene, headmaster of Winchester College (d. by 1484), fol. i.

Emden, 816; Ker, *Medieval Libraries*, 141 n. 8.

171. **Magdalen College Lat. 100** *c.* 1210 Worcester

Psalter (365 × 215)

Fine historiated initials pasted on separate pieces of vellum, several removed. Fine arabesque and penwork initials. [Illus.]

Binding of pink leather on boards, saec. XIV (?). Fore-edge pattern.

Prov. Calendar of Worcester, fols. 3–8^{v}, suggests a date between St. Wulfstan's canonization in 1203 (in the original hand) and his translation in 1218 (added in a later hand). Prior R. (obits of his parents added, fols. 3^{v}, 6^{v}), saec. XIII. Given by John Stanley in 1395, fol. 220.

Boase, *English Art*, 290–1; L. M. Ayres, 'A Tanner manuscript in the Bodleian Library', *J.W.C.I.*, XXXII (1969), 49, pl. 8b, d; Exh., *English illuminated manuscripts 700–1500*, Brussels, Bibliothèque Royale, 1973, no. 42; Morgan, *Early Gothic*, cat. 49; 12, 16, 38 n. 81, 96, 97; ills. 163–6.

172. **Magdalen College Lat. 149** *c.* 1220
Alexander Nequam, Comm. on the Song of Songs (405 × 283)

Fine arabesque initials. Fine penwork initials. Good marginal drawings (mutilated), sketches.

Prov. Given by magister Johannes Mowere, (d. 1489), 'precium istius libri vi li. xiij s. iiij d.', fol. 193.

Emden, 1327.

173. **New College 7** *c.* 1220 English (?)

Bible (375 × 240)

Fine historiated and other initials. Fine penwork initials. Canon Tables, fols. 294–7. Marginal instructions in Old French for the illuminator, fols. 1–97, followed later by preliminary sketches for initials or circles in the margins. [Illus.]

Prov. Acquired from the executors of Johannes Green (possibly John Green, fellow 1401–6), 'pro iiij libris vi s. viii d', fol. 300. Given by William Westbury, fellow, headmaster of Eton College in 1443, fol. 4^{v}.

Emden, 819, 2021; L. Eleen, *The illustrations of the Pauline Epistles in French and English Bibles of the Twelfth and Thirteenth Centuries*, 1982, 60 n. 70, 63–5, 67–8, ills. 90–103.

174. **Lincoln College Lat. 16,** fols. 1–138 Saec. XIII$\frac{1}{4}$

Haimo, Comm. on the Apocalypse (315 × 207)

Fine historiated and other initials. Fine penwork initials. Cf. no. 282. [Illus.]

175. **Merton College 210** (N.2.8) Saec. XIII$\frac{1}{4}$
Minor Prophets with gloss (330 × 243)

Fine historiated and other initials. Good penwork initials. [Illus.]

Prov. Bought in 1482 for 6s 8d by Richard FitzJames,

warden 1483–1507, bp. of London (d. 1522), who gave it to the College in 1494, fols. ii^{v}, 92^{v}.

Emden, 691–2.

176. **Oriel College 77** Saec. XIII$\frac{1}{4}$
Bible (335 × 210)

Fine and other initials. Fine penwork initials. Canon Tables (incomplete).

Fragments of medieval cover and pastedowns preserved as fly-leaves.

Prov. Pledged in 1477 by masters Crosse and Kente in the Celton (?) chest, Oxford (?), fols. 360–1. Given in 1596 by Robert Pierpont, later first earl of Kingston (d. 1643), fol. vi^{v}.

177. **Merton College 325** (H.1.5) Saec. XIII$\frac{1}{4}$
Petrus de Riga, Aurora (226 × 152)

Fine arabesque initials.

Prov. Ex libris of St. Mary's, Bridlington, press-mark (littera G), fol i^{v}.

Ker, *Medieval Libraries*, 12.

178. **Exeter College 45** Saec. XIII$\frac{1}{4}$
Bible (337 × 220)

Good penwork initials.

Prov. William Elyot, master of Godshouse, Portsmouth (d. 1494), fol. 4.

Emden, 638.

179. **St. John's College 100** Saec. XIII$\frac{1}{4}$
Bible (255 × 160)

Historiated (one) and other initials. Penwork initials.

Prov. Richard Butler, archdeacon of Northampton (d. 1612); given in 1613 through the good offices of John Buckeridge, bp. of Rochester and Ely, fol. 2.

180. **Balliol College 177** Saec. XIII$\frac{1}{4}$
Bede, Comm. on the Acts, etc. (310 × 210)

Good arabesque initials. Good penwork initials.

Prov. William Gray, bp. of Ely 1454–78, fol. 1^{v}.

181. **Brasenose College 5** Saec. XIII$\frac{2}{4}$
Oxford (?)

Bible (216 × 131)

Good initials. Good penwork initials. Diagrams of the Holy Land and of Solomon's Temple, fols 442–3^{v}, probably an addition of saec. XIII ex. [Illus.]

Prov. Revision of the text suggests Oxford.

H. H. Glunz, *History of the Vulgate in England*, 1933, 270–2.

182. **Balliol College 150** Saec. XIII$\frac{2}{4}$
Buildwas (?)

Bernard of Clairvaux, Sermons (392 × 290)

Good border, historiated initial. Good and other arabesque initials (mutilated). Marginal sketches, fols. 59, 92^{v}, 148, saec. XIV (?). Diagrams [Illus.]

Prov. Ex libris of Buildwas abbey, Shropshire (partly erased), fol. 2^{v}. Robert Thwaites, master, dean of Auckland, co. Durham (d. 1458), fol. 2^{v}.

183. **Balliol College 148** Saec. XIII$\frac{2}{4}$
English (?)

Bernard of Clairvaux, Various works (252 × 165)

Good initials. Good penwork initials.

Prov. '. . . quem emi ego frater . . . 1416 (?)'. 'Iste est liber R.P.' (cf. Mynors, *Cat.*, xli). William Gray, bp. of Ely 1454–78, fol. 1^{v}.

184. **Christ Church 107** Saec. XIII$\frac{2}{4}$
Bible (220 × 162)

Good historiated and other initials (probably retouched). Penwork initials.

Original binding of white leather on boards.

Prov. Bought in 1462 by magister R. Wrangwysh (d. 1517) from John More, University stationer, 'pro pretio librarum V (?)', fol. 462. Thomas Benet, 1533, fol. 1^{v}.

Emden, 2093.

185. **Corpus Christi College 3** Saec. XIII$\frac{2}{4}$
Bible (217 × 150)

Good historiated and other initials. Good penwork initials.

Prov. M. Lepethayte, subdean of Lincoln (d. 1488), fol. i^{v}. Given by magister Lathbury, fellow of Magdalen College (d. after 1517).

Emden, 1105, 1196.

186. **St. John's College 48** Saec. XIII$\frac{2}{4}$
Bible (305 × 185)

Fine historiated initials. Fine penwork initials. [Illus.]

Prov. Ex libris of a house of Austin Friars, fol. 2, saec. XV. 'Henricus' (mutilated), fol. 319^{v}, saec. XIV. Given by magister William Balow, 1598, fol. 1.

187. **Brasenose College 3** Saec. XIII$\frac{2}{4}$
Bible (155 × 108)

Good penwork initials.

Prov. Given by Nicholas Wilton to William Richardson, 1616, fol. 2, and by Walter Chetwynd de Ingestre to Richard Harrison, canon of Lichfield, 1667, who gave it to the College, fol. 3^{v}.

188. **Balliol College 172** Saec. xiii$^{2}_{4}$
Ezechiel, Daniel with gloss (332 × 233)

Good initials. Good penwork initials.

Prov. William Gray, bp. of Ely 1454–78, fol. 2^{v}.

189. **Magdalen College Lat. 108** Saec. xiii$^{2}_{4}$
Bible (295 × 200)

Fine historiated initials (mutilated). Fine penwork initials. [Illus.]

190. **Merton College 7** (A.2.7) Saec. xiii$^{2}_{4}$
Bible (305 × 198)

Fine historiated and other initials, By two hands, the first close to William de Brailes, cf. nos. 199, 200. Fine penwork initials. [Illus.]

Prov. Bequeathed by John Warkworth, fellow, master of Peterhouse, Cambridge, 1475–1500, fol. 1.

Emden, 1992; Morgan, *Early Gothic*, 115.

191. **Merton College 120** (M.2.4) Saec. xiii$^{2}_{4}$
Peter Comestor, Historia scholastica (350 × 242)

Fine and other penwork initials.

Prov. Pledged in 1431 in the University chest, Oxford, by Thomas Chace, master of Balliol College (d. 1499). Given by John Burbach, fellow, rector of St. John the Evangelist's, London (d. 1451), fol. 1.

Emden, 380, 306.

192. **Balliol College 280A–B** Saec. xiii$^{2}_{4}$
Gregory the Great, Moralia in Job, vols. I, III (325 × 220)

Fine arabesque initials. Marginal drawings by two hands, A, fols. 35, 103^{v}, 105^{v}; B, 107^{v}, 112, saec. xiv (?). Initial added, fol. 189^{v}, saec. xiii/xiv (retouched).

N. R. Ker, 'The English Manuscripts of the Moralia of Gregory the Great', *Kunsthistorische Forschungen Otto Pächt zu ehren, eds.* A. Rosenauer, G. Weber, 1972, 79 nos. 63, 64, 81 n. 23, 85.

193. **Balliol College 351** Saec. xiii$^{2}_{4}$
Bible (incomplete) (288 × 190)

Good arabesque initials. Penwork initials.

Prov. Bequeathed by Dr. George Coningesby, antiquary, 1766, first pastedown.

194. **Trinity College 82** Saec. xiii^{1}
St. Osyth.

Psalter (65 × 48)

Penwork initials.

Prov. Liturgical evidence points to St. Osyth (Chich), Essex. Obit of Juliana subpriorissa, 26 Aug., saec. xiii/xiv.

195. **Balliol College 39–40** Saec. xiii^{1}
Buildwas

Alexander Nequam, Comm. on the Song of Songs (incomplete) (362–74 × 245–83)

Good initials (mutilated), in vol. I added to a MS. of Saec. xii^{2}.

Prov. Initials in the style of MSS. from Buildwas abbey, Shropshire. Cf. no. 182. MS. 40, pledged in the Turville chest at Oxford by magister Thomas Chace, 1424–5, fol. 118^{v}. William Gray, bp. of Ely 1454–78, fol. 2^{v}, MS. 39.

Emden, 380; Mynors, *Catalogue*, xvii.

196. **New College 27** Saec. xiii^{1}
Acts, Epistles, Apocalypse, with gloss (350 × 220)

Fine borders, initials. Good penwork initials. [Illus.]

Prov. Given by the founder, William of Wykeham, bp. of Winchester 1367–1404.

197. **Lincoln College Lat. 31,** fols. 114–214.
Saec. xiii^{1}

Bede, Comm. on Canonical Epistles (260 × 170)

Good penwork initials, flourishes added to a MS. of *c.* 1200.

Medieval binding of white leather on boards, repaired.

Prov. Given by Robert Flemmyng, dean of Lincoln (d. 1483).

Weiss, *Catalogues*, 346 n. 17; Emden, 700.

198. **New College 29** Saec. xiii^{1}
Rhabanus Maurus, Comm. on Genesis (340 × 230)

Initials. Penwork initials.

Prov. Given by Thomas Lustcyll to magister Johannes Gynnor who gave it to the College, 1492, fols. 1^{v}, 190.

199. **New College 322** *c.* 1240–50
Oxford

Psalter (327 × 258)

Fine miniatures, borders, historiated and other initials. Fine penwork borders, initials. By William de Brailes and his workshop. Cf. nos. 190, 200. [Illus.]

Prov. Possibly made for a member of the de Brewes family of Beeding, Sussex. Calendar with obits of Mary de Brewes, d. 1323, Hugh Lemarny, d. 1334, John of Wokingdone, d. 1335. Given in 1693 by magister Henry Howell of London, first pastedown (cf. Hunt).

Exh., *Illuminated Manuscripts*, 1908, no. 59, pl. 59; S. C. Cockerell, *The work of William de Brailes* (Roxburghe Club), 1930, 7–11, pls. ii–xi; J. F. Willard, 'Occupations of the Months in medieval calendars', *B.Q.R.*, vii (1932), 36, 38; H. Swarzenski, 'Unknown bible pictures by W. de Brailes', *Journal of Walters Art*

Gallery, I, 1938, 63; G. Pollard, 'William de Brailes', *B.L.R.*, v (1955), 86 n. 2, 90–1, pl. 27; Rickert, *Painting in Britain*, 104–5, 235 n. 55; Morgan, *Gothic* MSS., 234; D. M. Robb, *Art of the illuminated manuscript*, 1973, 228; Hunt, *Medieval library*, 339 n. 81; R. Marks, N. Morgan, *The Golden Age of English Manuscript Painting 1200–1500*, 1981, 11; Morgan, *Early Gothic*, no. 74; 14, 30, 78, 81, 109, 114, 115, 117, 118, 121–3; ills. 239, 240, 248–9, figs. 14, 15.

200. **Christ Church 105** 1230–60 Oxford

Bible (300 × 215)

Fine historiated initial by William de Brailes (cf. no. 199). Fine penwork initials. Miniature pasted on fol. 3v removed. [Illus.]

Prov. David Lewis, D.C.L. of All Souls College, Oxford (d. 1584), fol. 1. William Pritchard, canon of Sarum and St. Paul's, London (d. 1629).

Morgan, *Early Gothic*, 115.

201. **Corpus Christi College 2** Saec. XIII med. St. Albans

Bible (369 × 240)

Fine and other initials. Fine penwork initials. Marginal notes, rubrics and additions by Matthew Paris of St. Albans (d. 1259). Cf. no. 25. [Illus.]

Prov. John Tyson, rector of St. Nicholas', Worcester, 1532–44, fol. 3.

R. Vaughan, 'The writing of Matthew Paris', *Transactions of the Cambridge Bibliographical Society*, I (1949–53), 389, 391, pl. XVIIIa; *id.*, *Matthew Paris*, 1958, 245–7, pl. 17; Thomson, *St. Albans*, 31, 112.

202. **All Souls College 1** Saec. XIII med. St. Augustine's, Canterbury (?)

Bible (220 × 137)

Fine initial. Good penwork initials (unfinished).

Medieval binding of white leather on boards, rebacked. Fore-edge pattern.

Prov. Ex libris and pressmark of St. Augustine's, Canterbury, saec. XIV, fols. iiv, iii, and catalogue of saec. XV ex. Given by Jenkyn Bowen, B.A. from All Souls College in 1610–11, fol. 1.

James, *Ancient Libraries*, 515.

203. **New College 284** Saec. XIII med.

Aristotle, Metaphysics (325 × 220)

Fine border, historiated and other initials. Good penwork initials, marginal ornament. [Illus.]

Prov. Possibly a copy of an Oxford University exemplar (evidence of format and contents).

204. **Exeter College 32** Saec. XIII med.

Peter Lombard, Sentences with gloss (incomplete) (330 × 212)

Fine penwork borders (mutilated), initials.

Prov. Given by Nicholas Sandwich to William Reed, bp. of Chichester 1368–85, fol. 1.

Emden, 1559, 1639.

205. **New College 20** Saec. XIII med.

Bible (Wisdom Books) with gloss (325 × 212)

Fine historiated and other initials. Good penwork initials. Preliminary marginal drawings. [Illus.]

Prov. Given by the founder, William of Wykeham, bp. of Winchester 1367–1404.

206. **Keble College 80** Saec. XIII med.

Bible (Genesis–Luke) (169 × 115)

Good marginal drawings, fols. 71, 72, 157. Good penwork initials.

Prov. Rev. Christie Chetwynd Atkinson (d. 1911).

207. **Christ Church 111** Saec. XIII med.

Bible, with gloss (124 × 80)

Good historiated initials. Penwork initials.

Prov. G. Griffith, M.A., 1628, fol. ii.

208. **St. John's College 110** Saec. XIII med.

Bible (mutilated) (242 × 170)

Fine borders, initials. Fine penwork initials.

Prov. Edward Bernard, fellow, librarian, 1667, fol. 1.

209. **Trinity College 24** Saec. XIII med. English (?)

Bible (235 × 180)

Fine historiated initial. Fine penwork initials. [Illus.]

Prov. Given in 1640 by Robert Skinner, fellow, while bp. of Bristol 1637–41 (d. 1670), fol. iv.

210. **Christ Church 109** Saec. XIII med.

Bible (195 × 122)

Good historiated and other initials. Penwork initials. [Illus.]

Prov. Given to N. J. Foucault whose arms are on the binding, saec. XVIII in. In England by 1728, fol. 1. Samuel Rogers, 'alumnus et bibliothecarius', 1784, fol. 1.

211. **Balliol College 222** Saec. XIII med.

Willelmus de Montibus, Similitudinarius, etc. (342 × 228)

Penwork initials. Faint marginal drawings, fols. 33, 47v, 48, saec. XIII/XIV.

Prov. William Gray, bp. of Ely 1454–78, fol. 1v.

212. **University College 17** Saec. XIII med.
Gregory IX, Decretals (158 × 110)

Fine historiated initial. Arabesque and penwork initials. [Illus.]

213. **Magdalen College Lat. 101** Saec. XIII med.
Peter Lombard, Sentences (300 × 220)

Good penwork initials.

214. **New College 358** Saec. XIII med. St. Albans

Psalter (255 × 180)

Fine border, initials, line fillers. Fine penwork borders, initials. [Illus.]

Prov. Calendar of St. Albans abbey, fols. iv–ix, with the feast of St. Edmund, archbp. of Canterbury (canonized 1246), in the original hand, suggests a date *c.* 1250. Given by Anne, widow of Robert Wetherall, fellow, (1784–1814), rector of Newton Longville, first paste-down.

Exh., *Illuminated Manuscripts*, 1908, no. 53, pl. 50; F. Wormald, 'English Benedictine Calendars after A.D. 1100', *H.B.S.*, LXXVIII (1938), 31, 34–45; Hunt, *Medieval Library*, 340.

215. **All Souls College 6** *c.* 1250–60 Salisbury (?)

Psalter (Amesbury Psalter) (305 × 205)

Important miniatures. Fine historiated and other initials. Fine penwork initials, line fillers. By the 'Sarum Master' and his associates, responsible also for the Henry of Chichester Missal (Manchester, John Rylands Univ. Library Lat. 24), the Wilton Psalter (London, Royal College of Physicians MS. 409) and the William of Hales Bible of 1254 (Brit. Libr. MS. Royal 1.B.XII). Marginal drawings added fol. 13, saec. XIV in. [Illus.]

Prov. Probably written for the nun shown kneeling before the Virgin, fol. 4, and Christ, fol. 6. St. Melorius in the calendar and the litany suggests Amesbury abbey, Wilts. Annotations by John Grandisson, bp. of Exeter 1327–69. Bequeathed by Daniel Lysons of Bath, fellow, 1772, fol. 1.

Exh., *Illuminated Manuscripts*, 1908, no. 41, pl. 39; Millar, I, 53 ff., pls. 81–3; A. Hollaender, 'The Sarum illuminator and his School', *Wiltshire Archaeological and Natural History Magazine*, L (1943), 239–48, pls. 6–13; Brieger, *English Art*, 174, pls. 64a, 65b; Rickert, *Miniatura*, II, 6, 11, 17, 18, pl. 12; *id.*, *Painting in Britain*, 105–6, pl. 106; Randall, *Images*, 40; D. H. Turner, *Early Gothic Illuminated Manuscripts*, 1969, 22–3, pl. 10; Morgan, *Gothic MSS.*, 238; Exh., *English Illuminated Manuscripts 700–1500*, Brussels, Bibliothèque Royale, 1973, no. 51; R. Marks, N. Morgan, *The Golden Age of English Manuscript Painting 1200–1500*, 1981, 14, fig. v; Morgan, *Early Gothic*, 122, 141.

216. **Corpus Christi College 1** *c.* 1250–60
Bible (274 × 195)

Fine historiated and other initials in the style of the William of Devon Bible, Brit. Libr. Royal 1.D.I. By three hands, one probably that of the William of Devon painter, another perhaps the artist of the Windmill Psalter (New York, Morgan Library, M. 102). Marginal drolleries added in saec. XIV[1]. [Illus.]

L. Eleen, *The illustrations of the Pauline Epistles in French and English Bibles of the Twelfth and Thirteenth Centuries*, 1982, 70 n. 114.

216 *bis.* **New College 306** *c.* 1250–60
Ps.-Chrysotom, Opus imperfectum on St. Matthew (380 × 225)

Fine historiated initial. Good penwork initials. By the William of Devon painter (cf. no. 216). Made for a Dominican friar shown kneeling before the Virgin, fol. 3^v^. [Illus.]

Prov. Bought from the executors of Simon Islip, archbp. of Canterbury (d. 1366), by William Reed, fellow of Merton College, bp. of Chichester 1368–85, who gave it to New College, fol. 2.

R. T. Gunther, 'Early Science in Oxford, II', *Oxford Historical Society*, LXXVIII (1923), 281; Emden, 1007, 1559; Hunt, *Medieval library*, 321.

217. **Corpus Christi College 285** *c.* 1270–80
Psalter (362 × 285)

Fine miniatures (calendar pictures, marginal grotesques). Fine historiated and other initials. Good penwork initials, line fillers. Good borders added saec. XIV in. with perhaps some additional retouching. Fols. 50–69 are a replacement of saec. XVI. [Illus.]

Prov. Calendar points to diocese of Canterbury (Oblatio S. Mariae, 21 Nov.). Given by Christopher Wase, fellow (d. 1711), fol. 1.

W. O. Hassall, *How they lived*, 1962, pls. I, II, XIX; *Rural Life*, pls. 19a, 19b, 22a.

218. **Merton College 72** (I.1.2) Saec. XIII$\frac{3}{4}$
Thomas Aquinas, Summa theologiae (320 × 218)

Good historiated initial (cf. the William of Devon group, no. 216). Penwork initials. [Illus.]

219. **St. John's College 29** Saec. XIII$\frac{3}{4}$
Bible (330 × 210)

Fine historiated (one) and other initials. Fine penwork initials. [Illus.]

Prov. Given by Robert Dow, merchant taylor, 1600, fol. 1.

220. **St. John's College 4** Saec. XIII$\frac{3}{4}$
Bible (400 × 253)

Fine border. Fine historiated and other initials (partly defaced and mutilated). Fine penwork borders, initials. [Illus.]

Prov. Given in 1620 by John Buckeridge, bp. of Rochester and Ely (d. 1631), fol. 1.

221. **Balliol College 348** Saec. XIII$\frac{3}{4}$
English (?)

Bible (incomplete) (110 × 65)

Good and other initials (cropped in binding).

222. **Jesus College 55** Saec. XIII$\frac{3}{4}$
Bible (230 × 145)

Fine historiated and other initials. By the same illuminator and written by the same scribe as Bodl. MS. Bodley 748. Good penwork initials. Marginal drawings. [Illus.]

Prov. Made for a Franciscan patron (or house), cf. fol. 8.

223. **Queen's College 358** Saec. XIII$\frac{3}{4}$
Bible (140 × 100)

Fine historiated and other initials. Good penwork borders, initials. [Illus.]

Prov. Richard Davis de Sanford, 1693, fol. i.

224. **Oriel College 78** Saec. XIII$\frac{3}{4}$
Bible (223 × 135)

Good historiated (one) and other initials.

Prov. Book-plate of Edward, fifth Baron Leigh (1742–86), who bequeathed it to the College.

225. **New College 2** Saec. XIII$\frac{3}{4}$
Bible (305 × 200)

Good arabesque and penwork initials.

Prov. John Dobbys, fellow 1488, fol. 345^{v}.

Emden, 579.

226. **Oriel College 9** Saec. XIII$\frac{3}{4}$
Bible (353 × 235)

Fine borders, initials (mutilated). Fine penwork initials.

Prov. Given in 1478 by Thomas Hawkyns, provost (d. 1479), fol. ii^{v}.

Emden, 892.

227. **Balliol College 1** Saec. XIII$\frac{3}{4}$
Bible (264 × 184)

Fine initials. Good penwork initials.

Prov. Given by George Holland, M.A., fellow, 1596, fol. 393^{v}.

228. **All Souls College 2** Saec. XIII$\frac{3}{4}$
Bible (265 × 175)

Fine historiated and other initials (partly defaced). Penwork initials. [Illus.]

Binding of white leather on boards.

Prov. Pledged in the Rothbury chest, Oxford, in 1491 by John Dobbys, and in the Chichele chest in 1492 by Thomas Curteys, both fellows of New College, Oxford, fol. i^{v}. Given by William Derlyngton, fellow of New College, Oxford, vicar of East Greenwich (d. 1527), fols. 1^{v}, 313.

Emden, 530, 574, 579.

229. **Brasenose College 4** Saec. XIII$\frac{3}{4}$
Bible (158 × 119)

Good initials.

Contemporary binding of pink leather on boards.

Prov. *Ex libris* of Durham cathedral, saec. XIII. Letter-mark (A), saec. XIV–XV. Given by Johannes Daniel, armiger, fol. 4^{v}.

230. **Lincoln College Lat. 19** Saec. XIII$\frac{3}{4}$
Aristotle, Metaphysica nova (345 × 225)

Good penwork initials.

Medieval boards with white leather covering, repaired.

231. **Corpus Christi College 59,** fols. 5–58
Saec. XIII$\frac{3}{4}$

Alan of Lille, Anticlaudianus (204 × 145)

Good and other marginal drawings.

Prov. Bound with miscellaneous texts possibly compiled in saec. XIII by a master of Lanthony grammar school (evidence of contents). Given to Lanthony, Glos., by Walter Waldink of Staunton, Herefordshire, saec. XIII ex., fol. 3. Given to the College in 1619 by Henry Parry, fellow, fol. 3.

C. Brown, 'A thirteenth-century MS. from Llanthony Priory', *Speculum*, III (1928), 587–95; *id.*, *English lyrics of the thirteenth century*, 1960, xxvii, 114–19, 212; C. Meier, 'Die Rezeption des Anticlaudianus Alans von Lille in Textkommentierung und illustration', *Text und Bild: Aspekte des Zusammenwirkens zweier Künste in Mittelalter und früher Neuzeit, eds.* C. Meier, U. Ruberg, 1980, 499–505, 506, 513–4, 518, Abb. 52–59.

232. **Exeter College 33** Saec. XIII$\frac{3}{4}$
Alexander of Hales, Summa theologica (345 × 235)

Good initial.

233. **Oriel College 14** Saec. XIII$\frac{3}{4}$
Gratian, Decretum with gloss of Bartholomew of Brescia (343 × 282)

Fine penwork initials (mutilated).

Original cover and pastedowns preserved as fly-leaves.

234. **New College 65** *c.* 1280
Apocalypse. In French (228 × 160)

Fine miniatures, historiated initial. Fine penwork initials. [Illus.]

Prov. Probably made for Joanna (d. 1283), wife of Humphrey de Bohun, fol. 44.

Exh., *Illuminated MSS.*, 1908, 44–5, no. 90, pl. 75; M. R. James, *The Apocalypse in Art*, 1931, 14, no. 61; F. D. Klingender, 'St. Francis and the birds of the Apocalypse', *J.W.C.I.*, XVI (1953), 18, 19, pl. 5a; R. Freyhan, 'Joachism and the English Apocalypse', *J.W.C.I.*, XVIII (1955), 219, 221, 233, pl. 53c; Brieger, *English Art*, 218; *Rural Life*, pl. 3a; Exh., *English Illuminated Manuscripts*, Brussels 1973, no. 57; Hunt, *Medieval library*, 337.

235. **Corpus Christi College 50** Saec. XIII ex.
Gregory the Grêat, Moralia in Job (455 × 315)

Fine initial. Fine penwork initials. [Illus.]

Medieval binding of white leather on boards.

Prov. Given by the founder, Richard Foxe, bp. of Winchester 1501–28, fol. 1.

Emden, 717; N. R. Ker, 'The English Manuscripts of the Moralia of Gregory the Great', *Kunsthistorische Forschungen Otto Pächt zu ehren, eds.* A. Rosenauer, G. Weber, 1972, no. 41, 79, 84.

236. **Corpus Christi College 95,** fols. 1–58^v Saec. XIII ex.
Daniel of Morley, Philosophia universalia (225 × 185)

Good penwork borders, initials. Diagrams. Fine marginal drawing, saec. XIV^1, fol. 46, Cf. no. 131.

Medieval binding of pink leather on boards.

L. Thorndike, 'Daniel of Morley', *English Historical Review*, XXXVII (1922), 541; Destombes, *Mappemondes*, no. 40.21.

237. **University College 116** Saec. XIII ex.
Bible (part II) (340 × 245)

Fine penwork borders, initials. Marginal scribbes.

Medieval boards, rebacked. Fore-edge painting with leaf pattern.

Prov. List of sermon subjects for the use of an order of Preachers, p. 306. John Thorgott, saec. XVI, pp. 1, 396.

Hunt, *Manuscript*, 27.

238. **Brasenose College 2** Saec. XIII ex.
Bible (155 × 110)

Good initials. Good penwork borders, initials.

Prov. Given by Edmund Leigh, fellow 1611–41, fol. 7.

239. **Merton College 213** (N.2.10) Saec. XIII ex.
Gospels with gloss (355 × 224)

Fine penwork initials, marginal ornament.

Prov. Given by Richard FitzJames, warden 1483–1507, bp. of London (d. 1522), who bought it in 1485 for 27 s., fols. i^v, 1.

Emden, 691–2.

240. **New College 105** Saec. XIII ex.
Peter Lombard, Sentences (375 × 247)

Fine borders, initials. Penwork initials.

Prov. Arms of the Kings of England, fol. 1; Earl of Cornwall, fol. 3^v; Edward the Confessor, fol. 126; unidentified: argent 3 charges (buckles?), bordure sable charged with bezants, fol. 72^v; checky or and gules on a bend azure 3 horse shoes argent, fol. 177. Bought for the price of XX s' from the executors of magister Thomas Hussley by Robert Heete, fellow (d. 1433), fol. 1.

Emden, 901.

241. **Merton College 269** (F.1.4) Saec. XIII ex.
Aristotle, Metaphysics, with comm. of Averroes (300 × 210)

Fine historiated initials (mutilated), preliminary marginal sketches, also later marginal drawings and scribbles. Fine penwork initials. Line-fillers. [Illus.]

Prov. Cautio, partly erased, of 1388, fol. 286^v.

Powicke, *Medieval Books*, no. 391; N. R. Ker, 'The Books of Philosophy distributed at Merton College in 1372 and 1375', *Medieval Studies for J. A. W. Bennett, ed.* P. L. Heyworth, Oxford 1981, 353, 371, fig. 4c.

242. **Merton College 235** (H.2.2) Saec. XIII ex.
Bible (275 × 180)

Fine historiated and other initials. Marginal drolleries. Penwork initials. [Illus.]

243. **Merton College 2** (A.2.2) Saec. XIII ex.
Augustine, Homilies on St. John (380 × 265)

Fine penwork initials.

Contemporary binding of white leather on boards.

Prov. Given in 1493 by the executors of John Tailour, provost of Oriel College, chancellor of Exeter (d. 1492), fol. ii^v.

Emden, 1851; N. R. Ker, 'Oxford College libraries in the sixteenth century', *B.L.R.*, VI (1957–61), 473.

244. **Magdalen College Lat. 2** Saec. XIII ex.
Bible (160 × 100)

Fine and other penwork initials.

245. **St. John's College 193** Saec. XIII ex.
Bible (170 × 115)

Good penwork initials.

Prov. Given in 1598 by Richard Lateware, fellow (d. 1601), fol. 4ᵛ.

246. **Jesus College 91** Saec. XIII ex.
Bible (mutilated) (300 × 200)

Good and other penwork initials. Marginal sketch, fol. 318, saec. XIV².

247. **Balliol College 330** Saec. XIII ex.
English (?)

Bible (198 × 122)

Good initial. Penwork initials. Good marginal drawings, fols. 105ᵛ, 113ᵛ (retouched), saec. XIII–XIV.

Prov. Scribbled entries of members of the Lewes family, saec. XVII in. Bequeathed by Dr. George Coningesby, antiquary, 1766, first pastedown.

248. **Merton College 111** (L.2.5) Saec. XIII ex.
English (?)

Peter Lombard, Sentences, with gloss (370 × 247)

Fine historiated initials. Good penwork initials, line-fillers. [Illus.]

Binding of medieval boards, resewn.

Prov. Bequeathed by Robert Whyte, fellow, rector of St. Peter's, Cheapside, London (d. 1461), fol. iiᵛ.

Emden, 2043.

249. **New College 187**, fols. 1–324ᵛ Saec. XIII ex.
Innocent IV, Apparatus on the Decretals (410 × 260)

Border, fine historiated initials. Fine penwork initials. [Illus.]

Prov. Unidentified arms, f. 4; argent chevrony gules, and argent barry azure. Adam de Orleton, bp. of Hereford 1317–27, a tract written for him *c.* 1311 added fols. 327–33. Thomas Jackson, fellow 1496–1510, fol. 4.

Emden, 1011; Hunt, *Medieval library*, 331–2.

250. **Balliol College 116** Saec. XIII ex.
Eustratius, Comm. on Aristotle's Ethics (340 × 215)

Fine historiated (one retouched) and other initials. [Illus.]

Prov. Pledged, jointly with other fellows, by Thomas Lyndelove, 1375, 1379, fol. iᵛ.

Aristoteles Latinus, I, no. 348; Emden, 1191.

251. **Balliol College 252** Saec. XIII ex.
Aristotle, de animalibus (340 × 210)

Good penwork borders, initials.

252. **Christ Church 106** Saec. XIII²
Bible (290 × 170)

Good penwork initials.

Prov. R. Lowe, of Cawne, Wilts., armiger, 1610, fol. 7ᵛ.

253. **Corpus Christi College 275**, fols. 113–169
Saec. XIII²
English (?)

Isaac Judaeus, de febribus, with gloss (218 × 155)

Good initials. Penwork initials. (Cf. no. 887).

Prov. Bought in London by Hieronimus Gandusinus de Verona, physician, 26 Aug. 1475, fol. 168.

254. **St. John's College 16** Saec. XIII²
Joshua-Esther, Acts, with gloss (380 × 225)

Good penwork initials (unfinished).

Prov. Richard Butler, archdeacon of Northampton (d. 1612). Given through the good offices of John Buckeridge, bp. of Rochester and Ely, 1613, fol. 1.

255. **Merton College 278** (O.1.6) Saec. XIII²
Aristotle, de animalibus, etc. (305 × 215)

Fine and other penwork initials.

256. **Lincoln College Lat. 7** Saec. XIII²
Thomas Aquinas, de veritate (194 × 140)

Good initials.

Prov. Given by the founder, Richard Flemmyng, bp. of Lincoln (d. 1431).

Weiss, *Catalogues*, 352 (134) n. 148; Emden, 698.

257. **Balliol College 129,** fols. 5–235ᵛ
Saec. XIII² (?)

Seneca, Various works (350 × 230)

Penwork borders. Good penwork initials. Good marginal drawings by the scribe, some coloured, fols. 5–98ᵛ. Cf. no. 854.

Prov. Robert Caldwell 'manentis apud Newark accomodatus Fratri Thome Box . . .', fol. 1ᵛ, saec. XIV². College inventory-mark, saec. XIV ex., fol. 4ᵛ. Annotated by Richard Bole, fellow, archdeacon of Ely (d. 1477).

258 **New College 116** *c.* 1300–20 Oxford

Thomas Aquinas, Comm. on the Sentences (334 × 225)

Fine historiated initial. Good penwork initials. [Illus.]

Prov. Written for an English Dominican house (?). Identified by *pecia* marks as copied from an Oxford University exemplar (?).

J. Destrez, *La Pecia dans les manuscrits universitaires du XIIIe et du XIVe siècle*, 1935, 100, pls. 29, 30; G. Pollard, 'The University and the book trade in medieval Oxford', *Miscellanea Medievalia*, III (1964), 342.

259. **University College 100** *c.* 1300

Apocalypse. French prose version with gloss (180 × 133)

Good miniatures (mainly coloured drawings). Good penwork initials. [Illus.]

Medieval binding of white leather on boards, remains of chemise pasted down.

Prov. Given in 1669 by William Rogers of Painswyck, commoner, fols. 1, 98.

M. R. James, *The Apocalypse in Latin and French* (Roxburghe Club), 1922, 16, 25 ff., no. 62; *id.*, *The Apocalypse in art*, 1031, 14, no. 62; G. Bing, 'The Apocalypse Block Books and their Manuscript Models', *J.W.C.I.*, V (1942), 150 n. 3.

260. **New College 316** *c.* 1300

Bible (incomplete) (180 × 120)

Fine and good penwork initials.

Prov. Given by Robert Yonge, fellow, rector of St. Michael's, Crooked Lane, London (d. 1479), fol. 413^{v}.

Emden, 2138.

261. **Merton College 96** (I.3.1) *c.* 1300

Thomas Aquinas, de veritate (355 × 222)

Fine border, historiated initial. [Illus.]

Prov. Given by Roger Martival, chancellor of Oxford University 1293–5, bp. of Salisbury (d. 1330).

Emden, 1233.

262. **All Souls College 49** *c.* 1300

Justinian, Digestum vetus with gloss of Accursius (427 × 265)

Fine miniatures, fine and other initials. [Illus.]

White leather on boards, saec. XVI (?).

Prov. Bury St. Edmunds, pressmark 'L.289', saec. XIV, fol. 1.

263. **Merton College 56** (G.2.1) *c.* 1300

Gregory the Great, Moralia in Job (379 × 240)

Fine penwork borders, initials.

Prov. Bequeathed by Henry de la Wyle, fellow, chancellor of Salisbury (d. 1329), fol. ii^{v}.

Emden, 565–6; N. R. Ker, 'The English Manuscripts of the Moralia of Gregory the Great', *Kunsthistorische Forschungen Otto Pächt zu ehren, eds.* A. Rosenauer, G. Weber, 1972, 79, no. 42.

264. **St. John's College 178** Saec. XIV in. London

Scientific miscellanies. Bestiary, etc. (189 × 139)

Fine miniature by two hands. Good initials, diagrams. Penwork initials. [Illus.]

Prov. Calendar of St. Peter's, Westminster (feast of Relics, 16 July; translation of Edward, the Confessor, 13 Oct.). List of contents in a contemporary hand, saec. XIV, fol. 1.

M. R. James, *The Bestiary* (Roxburghe Club), 1928, 19, pl. 20; Saxl and Meier, III, 1, 413, 2, pl. XCII; Destombes, *Mappemondes*, nos. 27.11, 41.10; L. F. Sandler, 'A follower of Jean Pucelle in England', *Art Bulletin*, LII (1970), 365; L. F. Sandler, *The Psalter of Robert de Lisle in the British Library*, 1983, 17, fig. 9.

265. **University College 86** Saec. XIV in. Durham (?)

Gratian, Decretals (390 × 250)

Good miniature. Fine initials. Fine penwork initials. Marginal sketch, fol. 267. Fols. 189–243 an insertion of saec. XIII in. with fine penwork initials. [Illus.]

Medieval binding of white leather on boards.

Prov. Ex libris of Durham cathedral and pressmark 'Q', saec. XV, fol. 1.

266. **Jesus College 40** Saec. XIV in.

Psalter (235 × 150)

Fine borders, historiated and other initials. [Illus.]

Prov. Calendar suggests London diocese (two feasts of St. Erkenwald). B. Jones, 1626, fol. 1. Given by sir John Conway (d. 1641) (?).

T. J. Fordyce, Sir T. M. Knox, 'The Library of Jesus College, Oxford', *Oxford Bibliographical Society Proceedings*, V (1937), 59; L. M. C. Randall, 'The snail in Gothic marginal warfare', *Speculum*, XXXVII (1962), 367, figs. 272, 314; *id.*, *Images*, 40, fig. 314; H. Helsinger, 'Images on the Beatus page of some Medieval Psalters', *Art Bulletin*, LIII (1971), 174, fig. 13; Exh., *Medieval Art in East Anglia, 1300–1520*, Norwich Castle Museum, 1973, 11, no. 5; L. F. Sandler, *The Peterborough Psalter in Brussels*, 1974, 98–9, 143 n. 3.

267. **University College 120,** pp. 1–125 Saec. XIV in.

Bestiary (240 × 155)

Good miniatures (mainly coloured drawings). [Illus.]

Medieval binding of white leather.

M. R. James, *A Bestiary of the 12th century* (Roxburghe Club) 1928, 20, 26, 54, suppl., pl. 21; Hunt, *Manuscripts*, 17 n. 1; D. J. A. Ross, 'A lost painting of Henry III's palace at Westminster', *J.W.C.I.*, XVI (1953), 160, pl. 22d.

268. **St. John's College 1** Saec. XIV in. Reading

Augustine, Homilies on St. John's Gospel (465 × 295)

Fine borders, fine and other initials. Written by Thomas Chakenden, monk of Reading (cf. Lidell, 54). Initials, fols. 322–26, addition of *c.* 1350. [Illus.]

Medieval boards with later leather cover.

Prov. No. 13 in the list of books from Reading abbey added to MS. St. John's College 11, saec. XIV. Lection for the anniversary of Henry I, founder of Reading abbey, added fols. 322–26. Given by John Stonor of North Stoke, Oxon., 1609, fol. i^v.

J. R. Liddell, 'Some notes on the library of Reading Abbey', *B.Q.R.*, VIII (1935), 49–50, 54.

269. **New College 174,** fols. 210–35 Saec. XIV in.

Frederic II, Apparatus feodorum with gloss (415 × 235)

Fine miniature, border. Good and other initials. Good penwork initials. Added to a MS. of saec. XIII^2. Italian script and parchment. Gloss added in England. Cf. no. 670. [Illus.]

Prov. Thomas Bourton.

270. **Balliol College 114** Saec. XIV in.

Averroes, Comm. on Aristotle, Physics, etc. (358 × 242)

Fine historiated and other initials related to the Walter de Milmete group, cf. no. 283. [Illus.]

Prov. Pledged by magister Lawrence Thornhill in the Guildford chest, fol. 1^v, and several times in the Vaughan chest, Oxford, fol. 186, 1324–28, who bequeathed it to the College (d. 1369). 'Quod Chace', fol. 2 (chancellor of Oxford University, 1426). 'Quod Burnham', fol. 2^v (d. 1451). Geoffrey Wykes, fellow, fol. 185^v, saec. XIV ex.

Aristoteles Latinus, 402, no. 347; Mynors, *Catalogue*, xvii; Emden, 2109.

271. **Corpus Christi College 159** Saec. XIV in.

Peter Comestor, Historia scholastica (313 × 213)

Fine and other initials. Fine penwork initials.

Prov. Lanthony abbey, fol. 211. Given by Henry Parry, fellow, 1618, fol. 1.

272. **Balliol College 221** Saec. XIV in.

Peter Comestor, Historia scholastica, etc. (347 × 225)

Good initial. Penwork initials.

Prov. Pledged in the New University chest, Oxford, by John de Lullington in 1336–7, 1339, 1340–2, fol. 224^v.

Emden, 1177.

273. **Merton College 188** (M.1.3) Saec. XIV in. English (?)

Nicholas Trivet, Comm. on the Gospels (294 × 183)

Initials.

Prov. Bequeathed by John Reynham, fellow, chancellor of Oxford University 1361–3 (d. 1376). Deposited before 1385 by Fr. Nicholas de Stenington at St. Frideswide's priory, Oxford, as caution for a loan of a copy of Bede's *Historia ecclesiastica*, fol. i.

Emden, 1571, 1771.

274. **All Souls College 7** *c.* 1310–20 East Anglia

Psalter (395 × 253)

Fine borders, historiated and other initials. By the best hand of the Tiptoft Missal (New York, Pierpont Morgan Library M. 107). Line-fillers. [Illus.]

Prov. Possibly Ely (feast of St. Etheldreda, in red). Obit of Maria de Pakenham, 1361, fol. 4. John Young, armiger, 1633, fol. 1.

Exh., *Illuminated Manuscripts*, 1908, no. 64, pl. 56; Millar, II, 80, no. 215; Saunders, *English illumination* I, 100, 102, pls. 115–6; L. F. Sandler, 'Christian Hebraism and the Ramsey Abbey Psalter', *J.W.C.I.*, XXXV (1972), pls. 12a, b, c, 13b, 14b, 15c; *Medieval art in East Anglia, 1300–1520*, Norwich Castle Museum, 1973, no. 19; B. Watson, 'The artists of the Tiptoft Missal and the Court Style', *Scriptorium*, XXXIII (1979), 34, 36, 38–9, pl. 7b.

275. **University Archives A.1** Saec. $XIV\frac{1}{4}$ Oxford

Chancellor's book (315 × 200)

Fine miniature (coloured drawing, defaced), fol. 13^v; cf. no. 317. [Illus.]

S. Gibson, *Statuta antiqua Universitatis*, 1931, x–xiii, 601–3.

276. **New College 264** Saec. $XIV\frac{1}{4}$

Boethius, de consolatione philosophiae, with comm. and other works by William Wheatley (350 × 252)

Fine borders, initials, diagram. Good penwork initials.

Prov. Given by magister Nicholas de Sandwich to William Reed, fellow of Merton College, bp. of Chichester 1368–85, fol. 2.

R. T. Gunther, 'Early Science in Oxford, II' *Oxford Historical Society*, LXXVIII (1923), 382; R. S. Loomis, 'Was there a play on the martyrdom of Hugh of Lincoln?', *Modern Language Notes*, LXIX (1954), 33; Emden, 1639.

277. **Merton College 289** (O.2.5) Saec. xiv$\frac{1}{4}$
Robert Kilwardby, Comm. on Aristotle, Prior analytics. Robert Grosseteste, Comm. on Aristotle, Posterior analytics, etc. (353 × 223)

Fine historiated and other initials. Good penwork initials. [Illus.]

278. **New College 51** Saec. xiv$\frac{1}{4}$
John Chrysostom, Works (330 × 205)

Good initial. Penwork initials.

Prov. Magister John Monks, fellow, vicar of Adderbury, Oxon. (d. 1414), fol. 130^{v}.

279. **Exeter College 42** Saec. xiv$\frac{1}{4}$
Huguccio, Etymologion (incomplete) (328 × 206)

Fine historiated and other initials. Preliminary marginal sketches. Penwork initials. [Illus.]

280. **Oriel College 46**, fols. 109–211 *c.* 1320–30
Liber Custumarum. In Latin and French (345 × 242)

Fine borders, historiated and other initials. By two hands, the second the artist of Cambridge, St. John's MS. s. 30 (256). Marginal preliminary sketches. Scribal ornament: incipits, cadels. Marginal drawings, fols. 140^{v}, 147, etc. Part of a volume written *c.* 1321, the rest of which is London, Brit. Libr., MS. Cotton Claudius D.II, fols. 1–24, 30–40, 42–115, 124–35, 266–77 and Corporation of London, Records Office, Liber Custumarum, fols. ii, 1–84, 86–102, 173–86. [Illus.]

Prov. London, Guildhall. William Fleetwood (d. 1594). Francis Tate, Magdalen College scholar, member of the Middle Temple (d. 1616), fols. iii, 65.

N. R. Ker, 'Liber Custumarum and other Manuscripts formerly at the Guildhall', *The Guildhall Miscellany*, III (1954), 37–45, pl.; *id.*, *Medieval Libraries*, 126–7 n. 6; *id.*, *Medieval Manuscripts in British Libraries, I, London*, 1969, 20–1; J. J. G. Alexander, 'English early fourteenth-century illumination: recent acquisition', *B.L.R.*, V (1974), 73 n. 3.

281. **Exeter College 29** *c.* 1320–30
Oxford

Innocent IV, Apparatus on the Decretals (345 × 215)

Good penwork borders, initials. Written in several typical university hands close to John Wylliot, fellow of Merton College, chancellor of Oxford University in 1349.

Prov. John Northwode, chancellor of Oxford University 1345–48, fol. i^{v}. Given to New College, Oxford, in 1483 by William Elyot, master of Godshouse, Portsmouth, fol. 4.

Emden, 638, 1372.

282. **Lincoln College Lat. 16,** fols. 139–181^{v}
c. 1320–30

Apocalypse (second family) with gloss. In French (315 × 207)

Fine miniatures, initials. Good penwork borders, initials. Cf. no. 174. [Illus.]

L. Delisle, *L'Apocalypse en francais au XIIIe siècle*, 1901, 227 n. 4; M. R. James, *The Apocalypse in Art*, 1931, 14, no. 60; G. Bing, 'The Apocalypse Block Books and their Manuscript Models', *J.W.C.I.*, V (1942), 150 n.

282 *bis*. **Lincoln College Lat. 113** 1320–30
Oxford

Guy d'Evreux, Sermones dominicales (387 × 240)

Fine border, historiated initial. Good penwork initials. Pecia marks. [Illus.]

Prov. Pledged by Thomas Pawnton, fellow (d. 1492), in the Selton chest, Oxford, 1481; renewed four times 1481–90, fol. 218^{v}.

J. Destrez, *La Pecia dans les manuscrits universitaires du XIIIe et du XIVe siècle*, 1935, 46, 99–100, pls. 27, 28; Emden, 1439.

283. **Christ Church 92** 1326–7
London (?)

Walter de Milmete, de nobilitatibus, sapientiis et prudentiis regum (245 × 160)

Important miniatures. Fine historiated and other borders, historiated and other initials. By several hands, the first, fols. 1–5^{v}, 59–60, responsible also for the drawings, fols. 72–78^{v}. Two hands identifiable in the sister MS. Brit. Libr. Add. 47680. From the same workshop as Bodl. MSS. Douce 231 and e Mus. 60 [Colour frontispiece: Illus.]

Binding of pink leather on original boards, patterned velvet chemise, saec. XV.

Prov. Composed for and presented to Edward III, fol. 1. Royal arms of England and of France in the borders, *passim* (cf. James, lxiv–lxxii). Given by William Carpender, alumnus, rector of Staunton upon Wye, Herefords, 1707, fol. i.

Exh., *Illuminated Manuscripts*, 1908, 33–4, no. 69, pl. 61; M. R. James, *The treatise of Walter de Milemete* (Roxburghe Club), 1913; Millar, II, 21–3, pls. 51, 53; Rickert, *Painting in Britain*, 135, pl. 137; Randall, *Images*, figs. 165, 320, 452, 689, 694; Ker, *Medieval Manuscripts in British*

Libraries, I, London, 1969, 428–9; Morgan, *Gothic MSS.*, 242; Pächt, Alexander, 3, nos. 575, 577–80, 587, 610; J. J. G. Alexander, 'English early fourteenth-century illumination', *B.L.R.*, IX (1974), 74 n. 3; J. J. G. Alexander, 'Painting and manuscript illumination for Royal patrons in the later Middle Ages', *English Court Culture in the Later Middle Ages, eds.* V. J. Scattergood, J. W. Sherborne, 1983, 141 n. 1, 142, pl.

284. **Balliol College 232B** Saec. xiv$\frac{2}{4}$
Oxford (?)

Aristotle, Physics, etc. (410 × 235)

Fine borders, historiated and other initials (mutilated). Good marginal drawings (catchwords and other). [Illus.]

Prov. Arms of Moels, fols. 11^{v}, 42, and Pomeroy, fols. 17, 31. Henry de la Pomeroy married Joan de Moels *c.* 1328. Their son, John, rector of Aveton Giffard, given a dispensation to study at Oxford in 1349. John Burgh, fellow, 1445, fol. 15.

E. B. Powley, *The House of de la Pomerai*, 1944, 58; Emden, 310; T. Kurose, *Miniatures of the Goddess Fortune in Mediaeval Manuscripts*, 1977, pl. 143; Ker, *College Libraries*, 298 n. 27, 300 n. 35.

285. **Magdalen College Lat 155** Saec. xiv$\frac{2}{4}$
East Anglia

Durandus, Speculum iuris (405 × 255)

Fine borders, initials (mutilated). Penwork initials in two styles. Marginal and other drawings.

Medieval binding of brown leather on boards.

286. **New College 208** Saec. xiv$\frac{2}{4}$

Gratian, Decretum with gloss (360 × 210)

Fine and good miniatures, historiated and other initials. Italian script and parchment, saec. XII ex. [Illus.]

Prov. Pledged by John Wynford, possibly fellow 1389–97, fol. 311.

Emden, 2059.

287. **New College 188** Saec. xiv$\frac{2}{4}$

Innocent IV, Apparatus on the Decretals (380 × 235)

Good border, initials.

Prov. Richard Andrew, fellow, dean of York (d. 1477), fol. 3, who bought it in 1432.

Emden, 35.

288. **Exeter College 30** Saec. xiv$\frac{2}{4}$

Innocent IV, Apparatus on the Decretals (385 × 240)

Good historiated and other initials. Penwork initials. Mutilated.

289. **Balliol College 301** Saec. xiv$\frac{2}{4}$

Boniface VIII, Sext. Constitutiones provinciales (390 × 240)

Fine borders, fine and other initials. Fine penwork initials.

290. **New College 207** Saec. xiv$\frac{2}{4}$
English (?)

Boniface VIII, Sext with gloss (380 × 232)

Fine miniatures, borders, initials. Good penwork initials. [Illus.]

Prov. Partly erased *cautio* of Richard Vaughan, fol. 156^{v}; marginal annotations possibly made by him *c.* 1316–34.

Emden, 1942; L. Boyle, 'Summa summarum and some other English works on Canon Law', *Proceedings of the 2nd International Congress on Medieval Canon Law*, 1965, 411, 416.

291. **Merton College 320** (O.3.9) Saec. xiv$\frac{2}{4}$

Henry Bracton, de legibus et consuetudinibus Angliae (342 × 238)

Good initials. Penwork initials.

Prov. 'Thomas . . .' with motto 'Doe well and drede not', fol. 194^{v}. Bequeathed by Thomas Farnelow, fellow, chancellor of York (1379).

Emden, 668.

292. **Merton College 244** (B.1.3) Saec. xiv$\frac{2}{4}$

Porchetus Salvaticus, Victoria contra Judaeos (243 × 162)

Good border, initial. Penwork initials.

Prov. Given by Richard Wykyng, vicar of Westwell, Kent, *c.* 1349, fol. ii^{v}.

293. **St. John's College 108** Saec. xiv$\frac{2}{4}$

Gilbertus Anglicus, Compendium medicinae (240 × 150)

Good border, historiated initial. Fine penwork borders, initials (unfinished).

Prov. Sir William Paddy, commoner, physician to James I, 1634, fol. 5.

294. **All Souls College 44** Saec. xiv$\frac{2}{4}$

Statutes of England. In Latin and French (102 × 63)

Border (mutilated), initials (unfinished). Good penwork borders, initials.

Prov. Narcissus Luttrell, 1693, fol. v^{v}.

295. **Lincoln College Lat. 5,** fols. 30–171
Saec. xiv$\frac{2}{4}$

Thomas Aquinas, Comm. on Sentences, Book II (305 × 215)

Good border, initial. Penwork initials.

Prov. Given by Thomas Gascoigne, chancellor of Oxford University (d. 1458).

Weiss, *Catalogues*, 345, 351, (120) n. 136; W. A. Pronger, 'Thomas Gascoigne', *English Historical Review*, LIII (1938), 618–9; Emden, 747.

296. **Merton College 33–4** (A.3.7–8) Saec. xiv$^{2}_{4}$
Augustine, Comm. on the Psalms (350–3 × 230)

Good borders (mutilated), good and other initials. Penwork initials.

Vol. I, binding of white skin over boards, saec. XVI, English.

297. **Oriel College 13** Saec. xiv$^{2}_{4}$
Thomas Aquinas, Comm. on St. Luke and on St. John (312 × 270)

Fine borders, initials (mutilated). Fine penwork initials. Catchwords, line-fillers.

Original cover and pastedown preserved as fly-leaves.

Prov. Given by Henry Sampson, provost 1449–76 (d. 1482), fol. vi^{v}.

298. **Merton College 37** (B.3.1) Saec. xiv$^{2}_{4}$
Augustine, Misc. works (mutilated) (325 × 220)

Fine border, historiated initial. Good penwork initial. [Illus.]

Prov. Pledged in 1350 by Richard Elyndon, fellow, rector of Lapworth (d. 1367), fol. 389^{v}.

Emden, 634.

299. **All Souls College 50** Saec. xiv$^{2}_{4}$
Justinian, Codex with gloss of Accursius (417 × 252)

Fine miniatures, borders, initials. Related in style to Christ Church, Oxford, MS. 92, Walter de Milmete (no. 283). French script. [Illus.]

Medieval binding of white leather on boards, repaired.

Prov. Given by John Southam, fellow, archdeacon of Oxford (d. 1441), fol. 6^{v}.

Exh., *Oxford College Libraries in 1556*, Bodleian Library, Oxford, 1956, no. 87.

300. **Magdalen College Lat. 209** Saec. xiv$^{2}_{4}$
Thomas Aquinas, Comm. on Aristotle, Posterior analytics. Aegidius de Columna, Comm. on Aristotle, de generatione (360 × 253)

Penwork borders, initials. Marginal scribbles.

301. **Merton College 22** (B.2.10), fols. 2–60
Saec. xiv^{1}

Augustine, de Trinitate (334 × 220)

Good penwork borders, initials.

Binding of white leather on boards paid for by John Burbach, fellow, rector of St. John the Evangelist's, London (d. 1451).

Prov. Given by John Reynham, fellow, chancellor of Oxford University 1361–3 (d. 1376), fol. 2.

Powicke, *Medieval Books*, 41; Emden, 1570.

302. **New College 23** Saec. xiv^{1}
Ezekiel, Daniel with gloss (310 × 225)

Fine penwork initials.

303. **Queen's College 389,** fols. 244–5 (fragment)
Saec. xiv^{1}
English (?)

Unidentified tracts, de paenitentia, etc. (approx. 267 × 168)

Historiated initials (rubbed). Penwork borders, initials. [Illus.]

304. **University College 61,** fols. 1–23^{v} Saec. xiv^{1}
Valerius, Epistola ad Rufinum, etc. (255 × 185)

Border, initial.

Medieval binding of brown leather on boards.

Prov. Richard Gosmore, fellow of Magdalen College (d. 1547); Thomas Broune, Christ Church scholar, vicar of Kingsclere (d. 1587), fol. i^{v}. Given by Thomas Walker, master (d. 1665), fol. 2.

Hunt, *Manuscript*, 23, 24; Emden, 794.

305. **New College 283** Saec. xiv med.
Aristotle, Tabula materiarum (280 × 160)

Border, initial. Good penwork initials.

306. **Exeter College 46** Saec. xiv med.
Psalter (incomplete) (295 × 195)

Fine miniatures (mutilated), borders. Fine borders, fine and other initials, line-fillers, by a different hand. [Illus.]

Fore-edge pattern.

Millar, II, 84; J. W. Thomas, *Exeter College, Oxford*, 1982, 8, pl.

307. **St. John's College 76** Saec. xiv med.
Guillelmus Parmensis, Chirurgia (285 × 192)

Good diagrams (coloured drawings). Penwork initials. [Illus.]

308. **New College 183** Saec. xiv med.
Clementines with apparatus of Johannes Andreae and of Guillelmus de Monte Laudano (370 × 197)

Good initials. Penwork initials, marginal drawing, fol. 195^{v}.

309. **Merton College 262** (C.2.13) Saec. xiv med.
John Gaddesdon, Rosa medicinae (241 × 168)

Good border, initial.

Medieval binding of white leather on boards.

Prov. Bequeathed by Robert Stonham, fellow (d. 1409), fol. ii^{v}.

Emden, 1790.

310. **Merton College 258** (C.2.9) Saec. xiv med.
Petrus de Crescentiis, Liber ruralium commodorum (254 × 182)

Good border, initial. Good penwork borders, initials (mutilated).

311. **Corpus Christi College 70** Saec. xiv med.
Gesellius de Cassaniis, Comm. on the Clementines, etc. (440 × 285)

Fine historiated and other initials by two hands added to a MS. of saec. xiv in. from South France with good penwork initials (mutilated). [Illus.]

312. **Balliol College 296** Saec. xiv med.
Abelard, Theologia 'Scholarium', etc. (410 × 280)

Fine borders, initials (mutilated). Good penwork initials.

Prov. Jacobus Stuartus, fol. i^{v} (added leaf), saec. xvii.

313. **Merton College 147** (H.1.4) Saec. xiv med.
English (?)
Nicholas Gorham, Distinctiones (230 × 162)

Initial. Penwork borders, initials.

314. **Exeter College 22** Saec. xiv med.
Johannes Lector (de Friburgo), Summa confessorum (320 × 195)

Fine borders, historiated and other initials (rubbed).

315. **New College 134** *c.* 1350
Thomas Bradwardine, de causa Dei. Thomas Buckingham, Quaestiones (340 × 222)

Border, good and other initials. Penwork initials.

Prov. William Reed, fellow of Merton College, bp. of Chichester 1368–85, fol. 2^{v}.

R. T. Gunther, 'Early Science in Oxford, II', *Oxford Historical Society*, lxxviii (1923), 382; Powicke, *Medieval Books*, 30; Emden, 1559; Hunt, *Medieval library*, 321 n. 19.

316. **Exeter College 47** Before 1373
Psalter (Bohun Psalter) (280 × 190)

Important miniatures. Fine historiated borders, historiated and other initials. By several different hands. [Illus.]

Original boards under patterned velvet cover.

Prov. Arms of England and Bohun in the borders. Written for Humphrey de Bohun, 7th earl of Hereford ('servus tuus Humfridus' in the prayers, fols. 118–122; d. 1373). Elizabeth, queen of Henry VII (d. 1503); Katherine of Aragon, queen of Henry VIII (div. 1531), fol. 1^{v}. Notes on the Royal family in the calendar end in 1511. Given by sir William Petre *c.* 1556, fol. ii.

Exh., *Illuminated Manuscripts*, 1908, no. 73, pl. 64; Millar, ii, 26, pls. 68–9; M. R. James, *The Bohun Manuscripts*, 1936, 5–22, pls. i-xxii; F. Harrison, *English Manuscripts of the 14th century*, 1937, 34, pl. 24; Rickert, *Painting in Britain*, 243–4 ns. 10, 12, 14; Randall, *Images*, 40; P. d'Ancona, E. Aeschlimann, *The Art of Illumination*. 1969, 33, pl. 99; Morgan, *Gothic MSS.*, 243; J. J. G. Alexander, 'Painting and manuscript illumination for Royal patrons in the later Middle Ages', *English Court Culture in the Later Middle Ages, eds.* V. Scattergood, J. W. Sherborne, 1983, 148.

317. **University Archives A.1,** fols. 17–28 1375
Oxford
Chancellor's book. Privileges of the University of Oxford granted by Edward III (315 × 200)

Good border, historiated initials. Cf. no. 275. [Illus.]

S. Gibson, *Statuta antiqua Universitatis*, 1931, x–xiii, 601–3.

318. **St. John's College 65** Saec. xiv$\frac{3}{4}$
Richard Fitzralph, Sermons (285 × 180)

Initial. Penwork initials. Lombards including human heads.

Prov. Dominican convent, Warwick: 'Liber fratris Nicholai Stremer quem partem emit et partem mendicavit a domino Johanne Savell monacho O.S.B., anno gracie 1479 tunc diaconus et curatus Oxon', fol. 203^{v}. Henry Savile (d. 1617). Sir William Paddy, commoner, physician to James I (d. 1634), fol. 1.

Ker, *Medieval Libraries*, 313; A. G. Watson, *The Manuscripts of Henry Savile of Banke*, 1969, no. 23.

319. **Oriel College 17** Saec. xiv$\frac{3}{4}$
Thomas Cobham, de paenitentia. Other miscellaneous texts (350 × 218)

Fine initials. Marginal drawings, fols. 1^{v}–2. Fine penwork initials.

Medieval cover and pastedowns preserved as fly-leaves.

320. **Oriel College 45** Saec. xiv$\frac{3}{4}$
Nicholas de Lyra, Comm. on the New Testament (475 × 280)

Good border, initial. Good penwork borders, initials.

Fragments of medieval cover and pastedowns preserved as fly-leaves.

Prov. Notes by Thomas Gascoigne, chancellor of Oxford University (d. 1458).

W. A. Pronger, 'Thomas Gascoigne', *English Historical Review*, LIII (1938), 618 n. 5; Emden, 746.

321. **Balliol College 95** Saec. XIV$\frac{3}{4}$
Walter Burley, Comm. on Aristotle, Ethics and Politics (incomplete) (400 × 280)

Fine border, initials. Fine penwork borders, initials.

Prov. Pledged in the Chichester chest, Oxford, by M. Walter 'M. . . .', 1405, fol. i^{v}. Bequeathed by John Malvern, chaplain of the College, canon of St. Paul's, London (d. 1422), fol. i^{v}.

Emden, 1211; C. Martin, 'Walter Burley', *Oxford Studies presented to Daniel Callus* (Oxford Historical Society, N.S., XVI), 1964, 226 n. 1.

322. **Oriel College 57** Saec. XIV$\frac{3}{4}$
Walter Burley, Comm. on Aristotle, Ethics and Politics (360 × 250)

Good border (mutilated). Fine and other penwork borders, initials.

Medieval cover and pastedowns preserved as fly-leaves.

C. Martin, 'Walter Burley', *Oxford Studies presented to Daniel Callus* (Oxford Historical Society, N.S., XVI), 1964, 226 n. 6.

323. **New College 242** Saec. XIV$\frac{3}{4}$
Walter Burley, Comm. on Aristotle, Ethics and Politics (mutilated) (380 × 260)

Fine borders, historiated initial. Good penwork initials, ornamented catchwords. Written by Johannes Balne, fols. 171^{v}, 250. [Illus.]

Prov. Given by Thomas Chaundler, fellow, chancellor of Oxford University (d. 1490), with his arms, fols. 1^{v}, 2.

M. R. James, *The Chaundler MSS.* (Roxburghe Club), 1916, 2, 5 n.; Emden, 399; Weiss, *Humanism*, 133; C. Martin, 'Walter Burley', *Oxford Studies presented to Daniel Callus* (Oxford Historical Society, N.S., XVI), 1964, 220 n. 2, 226 n. 1.

324. **Exeter College 17** Saec. XIV$\frac{3}{4}$
Guillelmus de Monte Laudano, Apparatus on the Clementines (433 × 280)

Fine and other initials. Good penwork initials.

325. **Balliol College 101** Saec. XIV$\frac{3}{4}$
Albertus Magnus, de vegetabilibus (225 × 145)

Good historiated initial (coloured drawing). Penwork border, initials. [Illus.]

Prov. Pledged in 1409 by William Saxton, fellow, who bequeathed it to the College (d. 1424).

Emden, 1648–9; Mynors, *Catalogue*, xix.

326. **Balliol College 245** Saec. XIV$\frac{3}{4}$
Aristotle, de animalibus, etc. (350 × 275)

Border, initial. Penwork initials.

Prov. William Gray, bp. of Ely 1454–78, fol. 2^{v}.

327. **Balliol College 33** Saec. XIV$\frac{3}{4}$
William of Nottingham, Comm. on Clement of Lanthony, Unum ex quatuor (306 × 263)

Good borders, initials (mutilated).

Prov. Given by John Waltham, subdean of York, fellow (d. 1384), fol. 1. Pledged by William Saxton, Roger Gowsell, Robert Forester, in Rothbury chest, Oxford, between 1415–18, fol. 356^{v}. Notes by Thomas Gascoigne, chancellor of Oxford University (d. 1458).

B. Smalley, 'Which William of Nottingham', *Medieval and Renaissance Studies*, III (1954), 215–6; Mynors, *Catalogue*, xix.

328. **Queen's College 308** Saec. XIV$\frac{3}{4}$
Heptateuch (376 × 280)

Fine historiated (partly defaced) and other initials. Fine penwork initials with human heads. Good marginal sketches for borders, saec. XV med. [Illus.]

Contemporary binding of boards.

Prov. 'Johannes de Hinburi me fecit', fol. 1. Made at his expense at a Franciscan house (?), fol. 10^{v}.

329. **Magdalen College Lat. 32** *c.* 1380
John Dumbleton, Summa logicae (262 × 162)

Fine border, initials. Fine drawing added, fol. ii, saec. XV in. [Illus.]

Contemporary binding of white leather on boards.

Prov. Dr. John Dee (d. 1608).

James, *Dr. John Dee*, 22; Emden, 603.

330. **Magdalen College Lat. 205** *c.* 1380
Walter Burley, Comm. on Aristotle, Ethics and Politics (370 × 230)

Fine borders, initials. Good penwork initials.

331. **Corpus Christi College 123** *c.* 1380
Calendar. Astrological treatise, etc. (220 × 150)

Fine miniature (zodiac man). Same iconography and style as Bodl. MS. Ashmole 391, V (cf. Pächt & Alexander, III, no. 678). Good and other initials. Diagrams.

Prov. Made for Joan, princess of Wales (d. 1385).

L. MacKinney, *Medical illustrations in medieval MSS.* (Wellcome Historical Medical Library, new series, v), 1965, 161.

332. **Trinity College 8** Before 1388
Missal. Use of Sarum (265 × 180)

Fine borders, historiated and other initials (mutilated). Fine penwork initials. [Illus.]

Prov. Note recording the birth in 1388 of Eleanor, first daughter of William De Bello Campo (Beauchamp) of Bergaveney, Herefordshire, fol. 4^v. Other notes relating to his family, 1397, 1403, fols. 5^v, 6^v. Requests for prayers for Walter Bates, scholar of Lincoln College (?) (d. 1479), fols. 4, 9^v.

M. Rickert, *The reconstructed Carmelite Missal*, 1952, 79 n. 1, pl. 45a; Morgan, *Gothic MSS.*, 243 n. 30.

333. **Corpus Christi College 394** 1398
Missal (Lapworth Missal). Use of Sarum (425 × 275)

Important miniature, marginal drawing. By the so-called 'Johannes associate' of the Herman Scheerre workshop. Fine borders, fine and other initials. Good penwork initials. Written in 1398, fol. 257 (erased inscription). [Illus.]

Prov. Given to the parish church of Lapworth, Warwicks., by Thomas Asheby (d. 1443). Given to the College by Henry Parry, fellow, 1618.

M. Rickert, *The reconstructed Carmelite Missal*, 1952, 93–4, pl. L; *id.*, *Miniatura*, II, 1959, 24–5, tav. 44; Exh., *Europäische Kunst um 1400* (8th Council of Europe Exhibition), Vienna, 1962, no. 145; Rickert, *Painting in Britain*, 156, 167, 171, 175, 245 n. 32, pl. 168; Exh., *English illuminated manuscripts*, Brussels, 1973, no. 72, pl. 35; G. M. Spriggs, 'The Nevill Hours and the school of Herman Scheerre', *J.W.C.I.* XXXVII (1974), 123–6; D. H. Turner, 'The Wyndham Payne Crucifixion', *British Library Journal*, II (1976), 11.

334. **St. John's College 130** Saec. XIV ex.
St Albans (?)
Hugo de Folieto, Super novem beneficiis religionis, etc. Bernard of Clairvaux, Meditationes, etc. (mutilated) (220 × 147)

Good borders, initials. Fine penwork borders, initials.

Prov. Given by fr. Henry Wylum to fr. John Bywell by permission of abbot John (Whethamstede (?)), fol. 187^v (erased *ex libris*), saec. XV^1. Sir William Paddy, commoner, physician to James I (d. 1634), fol. iv^v.

Ker, *Medieval Libraries*, 168, 302.

335. **Merton College 54** (B.3.5) Saec. XIV ex.
Bernard of Clairvaux, Misc. works (327 × 240)

Fine penwork borders, initials.

336. **Magdalen College Lat. 98** Saec. XIV ex.
Robert Grosseteste, Dicta (320 × 215)

Fine borders, initials. Penwork initials.

Prov. Bequeathed by William Wagge, rector of St. Mary's, Winchester (d. 1469), to William Ive, chancellor of Sarum (d. 1486), who gave it to the College, fol. 2.

Emden, 1008, 1954.

337. **Corpus Christi College 89** Saec. XIV ex.
Ranulf Higden, Polychronicon (330 × 223)

Good borders, initials (mutilated). Good penwork initials. *Mappa mundi.* [Illus.]

Prov. *Ex libris* of St. Peter's abbey, Gloucester, p. 12. Johannes Rheseus, 1539, p. 23. Notes by sir John Prise (d. 1555), pp. 12, 27, etc.

N. R. Ker, 'Sir John Prise', *The Library*, X (1956), 7, 16, 23; Destombes, *Mappemondes*, no. 47.13; J. W. Taylor, *The Universal Chronicle of Ranulf Higden*, 1966, 67–8, 106, 124, 157.

338. **Magdalen College Lat. 181** Saec. XIV ex.
Ranulf Higden, Polychronicon (340 × 220)

Good penwork initials, marginal scribbles, diagram. Fine drawing added, fol. i, saec. XV^2. [Illus.]

J. W. Taylor, *The Universal Chronicle of Ranulf Higden*, 1966, 98, 157.

339. **New College 152** Saec. XIV ex.
Ranulf Higden, Polychronicon with continuation to 1377 (375 × 243)

Fine borders. Fine and other initials. [Illus.]

Prov. Given by William North, fellow, vicar of North Curry, Somerset, etc. (d. 1462), fol. 165^v.

Emden, 1368; J. W. Taylor, 'The development of the Polychronicon continuation', *English Historical Review*, LXXVI (1961), 32 n. 2, 36; *id.*, *The Universal Chronicle of Ranulf Higden*, 1966, i, 2, 108, 123, 151, 157, 180.

340. **Merton College 10** (P.1.3) Saec. XIV ex.
Robert Grosseteste, Dicta. Chrysostom, Comm. on St. Matthew. Gregory the Great, Homilies (345 × 240)

Good border, initial. Good penwork initials.

341. **Lincoln College, Lat. 107** Saec. XIV ex.
Ranulf Higden, Polychronicon with continuation to 1377 (410 × 285)

Fine borders, initials. Good penwork initials. [Illus.]

Prov. Probably given by the founder, Richard Flemmyng, bp. of Lincoln (d. 1431).

Weiss, *Catalogues*, 349 (57) n. 82; J. W. Taylor, *The Universal Chronicle of Ranulf Higden*, 1966, 123, 151, 181.

342. **University College 177 A–B** Saec. xiv ex.
Ranulf Higden, Polychronicon (third version) (270 × 254)

Fine penwork borders, initials.

Contemporary binding of white leather on boards, chemise. Bookmarker, vol. 1.

Prov. Evidence for Barnwell priory, Cambridgeshire, from obits in the margin.

V. H. Galbraith, 'Extracts from the Historia aurea and a French Brut', *English Historical Review*, xliii (1928), 205 n. 1; Hunt, *Manuscript*, 26; Ker, *Medieval Libraries*, 7.

343. **Merton College 156** (L.3.8) Saec. xiv ex.
William of Nottingham, Comm. on the Gospels (375 × 264)

Fine border, initial. Good penwork initials.

Prov. Given by John of Bloxham, fellow, warden (d. 1387), fol. ii^{v}. Rebound at the expense of John Burbach, fellow, rector of St. John the Evangelist's, London (d. 1451), fol. i^{v}.

B. Smalley, 'Which William of Nottingham', *Mediaeval and Renaissance Studies*, iii (1954), 218; Emden, 204.

344. **Corpus Christi College 69** Saec. xiv ex.
John Gaddesdon, Rosa medicinae (325 × 214)

Fine border, historiated and other initials.

Emden, 739.

344 *bis*. **New College 178** Saec. xiv ex.
Bartolus de Sassoferato, Comm. on Digestum novum. On paper (410 × 280)

Good initials. Penwork initials. Marginal drawings. Paper manufactured in North Italy (watermarks C. M. Briquet, *Les filigranes*, 1968, iii, no. 7457; iv, no. 15820). Possibly written in Italy.

Prov. Cartouche 'Hovyngham', f. 230, perhaps John Hovyngham, archdeacon of Durham, who attended the Council of Constance in 1415 (d. 1417). Given by William Spekynton, fellow (d. 1509), fols. 1^{v}, 2.

Emden, App. 2184.

345. **St. John's College 14** Saec. xiv ex.
Peter of Herenthals, Comm. on the Psalter (365 × 230)

Borders, fine historiated initials. Penwork initials. [Illus.]

Prov. Ex libris of Durham cathedral, letter-mark 'G', saec. xv (?), fol. 1. Given by John Stonor of North Stoke, Oxon., saec. xvii in. (?).

346. **Merton College 214** (N.1.8) Saec. xiv ex.
Guillaume Peyraut, Sermones dominicales (290 × 200)

Fine borders, initials. Penwork initials. Written by Mitton, fols. 123^{v}, 293^{v}.

Prov. Given by Robert Elyot, fellow of All Souls College, vice-provost of Eton College (d. 1498), fol. iii^{v}.

Powicke, *Medieval Books*, 223 n., no. 1129; Emden, 638.

347. **Lincoln College Lat. 26** Saec. xiv ex.
Bernard of Clairvaux, Comm. on the Song of Songs (250 × 187)

Fine border, initial, good pen flourishes added to a MS. of saec. xii med.

Prov. Given by John Duffeld before 1474 with a request for prayers for Thomas Duffeld, chancellor of Lincoln cathedral (d. 1423), fols. 1, 2^{v}.

Weiss, *Catalogues*, 348 (20) n. 42; Emden, 600.

348. **All Souls College 302** Saec. xiv ex.
Missal (incomplete) (389 × 276)

Good initials. Border, initial, fol. 124, added saec. xv$\frac{2}{4}$. Penwork initials.

Prov. 'Habeatur in memoria Henricus Cantuariensis Archiepiscopus (Henry Chichele, d. 1443), fol. 186^{v}. Obit of William Kele, warden (d. 1459), fol. iv^{v}. Garbrand, Oxford bookseller, *c.* 1549/50, fol. 1. Found at Tusmore, Oxfordshire, in 1841. Bought back from the executors of sir Henry Dryden, 1899.

'A medieval service book from All Souls', *B.L.R.* iii (1950–1), 120.

349. **University College 178** Saec. xiv ex.
Missal. Use of Sarum (fragment) (395 × 262)

Fine border, initials. [Illus.]

Prov. University College chapel. Obits in the calendar: magister Walter Skirlaw, bp. of Durham, 1406; Henry Percy, 1455. Note on the dedication of the Chapel to St. Cuthbert, 1476.

C. Wordsworth, 'The Ancient Kalendar of the University of Oxford', *Oxford Historical Society*, xlv (1903–4), 38; Hunt, *Manuscript*, 17 n. 2, 19.

350. **New College 90** Saec. xiv ex.
Ralph Fitzralph, Sermons (390 × 252)

Fine borders, initials. Penwork initials. [Illus.]

Prov. Arms of Richard Scrope, archbp. of York, 1382–1405, fols. 1, 125. William Aiscough, bp. of Salisbury, 1438–50, fol. 1^{v}.

Emden, 16.

351. **Merton College 310** (H.3.10) Saec. XIV ex.
Robert Grosseteste, Summa philosophiae (258 × 176)

Fine border, historiated initial. Good penwork initials. [Illus.]

352. **Corpus Christi College 201** Saec. XIV ex.
William Langland, Piers Plowman ('B' text). In English (294 × 174)

Good border, historiated initial. Marginal sketches, fols. 11, 93. Penwork initials. [Illus.]

Prov. William Fullman, fellow, rector of Meysey Hampton, Glos. (d. 1688), fol. 1.

G. Kane, E. T. Donaldson, *Piers Plowman—B version*, 1975, 8.

353. **Worcester College 2.1** Saec. XIV ex.
Chandos Herald, Poem on the Life of the Black Prince. In Anglo-Norman (235 × 140)

Good border, initial. Penwork initials.

Prov. William Le Neve, Mowbray herald, later Clarencieux King of arms (d. 1661), fols. 1, 18v. Given by Dr. George Clark, fellow of All Souls (d. 1736).

M. K. Pope, E. C. Lodge, *Life of the Black Prince by the Herald of Sir John Chandos edited from the manuscript in Worcester College*, 1910; D. B. Tyson, *La vie du Prince Noir by Chandos Herald. Edited from the manuscript in the University of London Library, Beihefte zur Zeitschrift für Romanische Philologie*, 147, Tübingen, 1975, 1–14 and *passim*.

354. **Balliol College 158** Saec. XIV ex.
Constitutiones provinciales, with gloss, etc (287 × 183)

Good border, initials (mutilated). Penwork initials.

Prov. Richard Bole, fellow, archdeacon of Ely (d. 1477), fol. 1v.

355. **Balliol College 225** Saec. XIV ex.
St. Bridget, Revelations (303 × 222)

Good borders, initials. Penwork initials.

Prov. Notes and corrections by Thomas Gascoigne, chancellor of Oxford University (d. 1458), *passim*.

356. **Balliol College 86** Saec. XIV ex.
Guillaume Peyraut, de vitiis et virtutibus. Durandus, Rationale, etc. (375 × 248)

Good borders, initials. Penwork initials.

Prov. William Gray, bp. of Ely 1454–78, fol. 1v.

357. **Queen's College 307** Saec. XIV ex.
Ranulf Higden, Polychronicon (375 × 198)

Good penwork borders, initials. Diagram.

Prov. Given by abbot Thomas Hunden (d. 1419) to St. Augustine's, Canterbury. Shelfmark 'D.G.', saec. XV. Given by Dr. John Dee to magister Blomfeld in 1561. Bequeathed to the College by Henry Airey, provost (d. 1616), fol. 2.

James, *Ancient Libraries*, 297, no. 936; *id.*, *Dr. John Dee*, 28.

358. **Pembroke College 2,** fols. 11–359 Saec. XIV ex.
John Mirfield, Breviarium Bartholomei, etc. (375 × 240)

Fine border (rubbed), one historiated and other initials. Marginal diagrams, fol. 17. Drawing, fol. 359, saec. XV–XVI (?). [Illus.]

Original boards with white leather cover, metal bosses, straps and pins, remains of a chemise, saec. XV. Repaired.

Prov. Arms of Abingdon abbey, fol. 11. Bound with a calendar calculated for the meridian at Oxford 1387 by magister Walter Elvesden *c.* 1400. Given to All Souls College by Richard Bartlett, fellow (d. 1566), erased inscription fol. 12.

L. MacKinney, *Medical illustrations in medieval MSS.* (Publications of the Wellcome Historical Medical Library, new series vol. V), 1965, 162.

359. **St. John's College 2** Saec. XIV ex.
William of Nottingham, Comm. on the Gospels (incomplete) (435 × 278)

Fine borders, initials. Fine penwork borders, initials, line-fillers. [Illus.].

Prov. Given by John Stonor of North Stoke, Oxon., in 1609, first pastedown.

B. Smalley, 'Which William of Nottingham', *Mediaeval and Renaissance Studies*, III (1954), 219.

360. **Exeter College 6** Saec. XIV ex.
Ps.-Chrysostom, Opus imperfectum on St. Matthew (233 × 162)

Good border, initial.

Prov. Bequeathed by John Alward, fellow, rector of Stoke Bruere (d. 1458).

361. **Keble College 47** Saec. XIV ex.
Book of Hours. Use of Sarum (222 × 150)

Fine and good miniatures, borders, historiated and other initials. Line-fillers. By three hands. Hand II, fols. 9, 10v, in the same as hand C1 of Brit. Libr. MS. Add. 29704–5, the Carmelite Missal. Hand III is related to Hand D of Bodl. MS. Laud Misc. 165. (Cf. Pächt, Alexander, 3, no. 739). [Illus.]

Prov. Shield with arms showing a swan statant wings erect argent facing to sinister on a gray-green ground, fol. 57, suggest it may have been made for a member of the Bohun family. Sir Thomas Brooke, Huddersfield (d. 1908). Bequeathed by his brother, Canon C. E. Brooke (d. 1911).

K. L. Scott, 'John Lydgate's Lives of Saints Edmund and Fremund, a medieval manuscript in Arundel Castle', *Viator*, XIII (1982), 347 n. 50.

362. **Trinity College 77** Saec. XIV ex.
Lectionary. Use of Sarum (288 × 190)

Fine borders, historiated and other initials. By the same hand as Bodl. MS. Bodley 581 (cf. Pächt, Alexander, 3, no. 673). Fine penwork initials. [Illus.]

Proc. Given by the founder, sir Thomas Pope (d. 1559), fols. 1, 136^{v}.

363. **All Souls College 86** Saec. XIV^{2}
Walter Burley, Comm. on Aristotle, Miscellaneous works (389 × 280)

Fine penwork borders, initials.

Prov. John Stokes, warden 1466–94, fol. 2.

364. **Magdalen College Lat. 134** Saec. XIV^{2}
Summa summarum (440 × 235)

Fine borders, initials (mutilated).

Prov. Arms of the Kings of Scotland and of the Butler or Manderville families, fol. 1. Caution (erased), 1411. Bought by Thomas Wyche, fellow of Oriel College, from John More, Oxford bookseller, on 21 Oct., 1457, fol. 302^{v}.

365. **Magdalen College Lat. 137** Saec. XIV^{2}
Bartolomaeus Anglicus, de proprietatibus rerum (370 × 260)

Fine initials. Good penwork initials. Written by Johannes de Mepeham, fol. 281.

366. **Corpus Christi College 144,** fols. 97–104^{v} Saec. XIV^{2}

Astronomical treatises (225 × 175)

Border (mutilated), good initial.

Prov. Given to Tynemouth priory by John Bamburgh, subprior, 1447, fols. 15^{v}, 26. Dr. John Dee (d. 1608). Bequeathed by Brian Twyne, fellow (d. 1644), fol. 1.

James, *Dr. John Dee*, 36.

367. **Corpus Christi College 118,** fols. 57–113^{v} Saec. XIV^{2}

Boethius, de arithmetica (210 × 122)

Penwork initials. Good diagrams. Cf. no. 35.

Prov. Robert Greene de Welbe, fol. 118^{v}, saec. XV (?). Dr. John Dee, cf. no. 35.

368. **Corpus Christi College 78,** fols. 3–214 Saec. XIV–XV
Brut Chronicle with continuation to 1397 (165 × 110)

Fine drawing. Good initial. Penwork initials. [Illus.]

Prov. Given by Henry Parry, fellow, 1619, fol. 1.

V. H. Galbraith, 'Extracts from the "Historia aurea" and a French "Brut"', *English Historical Review*, XLIII (1928). 206.

369. **Lincoln College Lat. 105** Saec. XIV–XV
Robert Grosseteste, Summa justitiae, etc. (407 × 280)

Good and other penwork initials.

Medieval binding of white leather on boards with chemise pasted down.

Prov. Thomas Gascoigne, chancellor of Oxford University (d. 1458). Robert Flemmyng, dean of Lincoln (d. 1483), first pastedown.

N. R. Ker, 'Chaining from a staple on the back cover', *B.L.R.* III (1950–1), 104; Emden, 700, 747; Weiss, *Humanism*, 103 n. 3.

370. **St. John's College 86** Saec. XIV–XV
John Arderne, Liber medicinarum (278 × 180)

Diagram. Good marginal and other drawings (partly coloured).

Prov. Given by Mr. Backter to William Stokes, surgeon, 1571, page 1. Sir William Paddy, commoner, physician to James I, 1634, fol. 1.

371. **Corpus Christi College 17** *c.* 1400
Psalter (350 × 227)

Fine borders added to a French MS. of saec. XIII med. (cf. no. 668). [Illus.]

Prov. Obits of John Malloc and of Nichola, his wife, added to the calendar, suggest possibly the Mallocs of Axmouth, Devon. George Lloyd of Whetenhurst, Glos., 'procurante magistro Fulman . . .', *i.e.* William Fullman, fellow, rector of Meysey Hampton, Glos., (d. 1688), fol. iv^{v}.

372. **Queen's College 383** *c.* 1400
John Mandeville, Travels. In Latin and English (135 × 90)

Good borders, initials. Penwork initials. Written by Edward Jenkyn, fol. 162^{v}.

Contemporary binding of white leather on boards.

Prov. Given in 1671 by Thomas Pargiter, Lincoln College D.D. (d. 1705), fols. i, ii^{v}.

J. W. Bennett, *The rediscovery of Sir John Mandeville*, 1954, 292.

373. **Merton College 215** (N.1.9) *c.* 1400
St. Bridget, Revelations (325 × 215)

Fine border, initials. Penwork initials.

374. **Magdalen College Lat. 208** *c.* 1400
Guy de Chauliac, de chirurgia (355 × 257)

Good initials (mutilated). Good penwork initials.

Binding of white skin over boards, repaired.

375. **Magdalen College Lat. 104** *c.* 1400
Johannes de Burgo, Pupilla oculi (360 × 230)

Good and other initials.

Contemporary binding of white skin on boards.

376. **Magdalen College Lat. 110** *c.* 1400
Johannes de Burgo, Pupilla oculi (285 × 200)

Fine borders, initials (mutilated). Good penwork initials.

Prov. Magister John Tailour, possibly fellow of Oriel College (d. 1492), fol. i^{v}. John Hyett, perhaps Magdalen Hall 1662, last pastedown.

377. **Lincoln College Lat. 108** *c.* 1400
Rhabanus Maurus, Comm. on St. Paul's Epistles (385 × 275)

Fine and other penwork initials.

Prov. Given by magister Thomas Heton, rector of Mancetter, Warwicks. (adm. 1432), fol. 1.

378. **All Souls College 37** *c.* 1400
Flores historiarum (375 × 245)

Good border, initial. Good penwork border, initials.

Prov. Bequeathed in 1438 by Henry Penwortham, registrar of the founder, Henry Chichele, archbp. of Canterbury 1414–43, fol. 1.

379. **Lincoln College Lat. 110** *c.* 1400
Robert Holcot, Comm. on the Book of Wisdom (375 × 252)

Fine border, initials. Good penwork initials. Fragment of an illuminated MS., saec. XIV ex., used as fly-leaves.

Medieval binding of white leather on boards, saec. XV.

Prov. Given by John Forest, dean of Wells (d. 1446).

Weiss, *Catalogues*, 348 (27) n. 50; N. R. Ker, 'Chaining from a staple on the back cover', *B.L.R.* III (1950–1), 104; Emden, 706.

380. **Lincoln College Lat. 78** *c.* 1400
William of Nottingham, Quaestiones (258 × 175)

Fine initial, border, Penwork initials.

Prov. Given by John Forest, dean of Wells (d. 1446), fol. 1^{v}.

Weiss, *Catalogues*, 348 (19) n. 40; Emden, 706; B. Smalley, 'Which William of Nottingham', *Medieval and Renaissance Studies*, III (1954), 226.

381. **Magdalen College Lat. 52** *c.* 1400
Richard Rolle, Comm. on the Psalter (220 × 140)

Border, initials.

382. **Exeter College 2** *c.* 1400
Sermons of the Order of Preachers (315 × 208)

Good borders, initials (mutilated). Penwork initials.

383. **All Souls College 42** *c.* 1400
Constitutiones provinciales Angliae, etc. with gloss of John Acton (285 × 202)

Good border, initial. Penwork initials.

Prov. Given by master Thomas Bisshop, rector of Goathurst, Somerset (adm. 1473), to master Richard Gaunt (d. by 1519) with the reversion to the College, fol. 316^{v}.

Emden, 194, 749.

384. **Balliol College 166B, 171** *c.* 1400
Nicholas de Lyra, Comm. on the Old Testament (incomplete) (450 × 330)

Fine borders, initials (mutilated). Good penwork initials.

Prov. Given by Richard Colling, fellow, vicar of St. Lawrence Jewry, London (d. 1438).

Emden, 466.

385. **Exeter College 134** *c.* 1400
Placita coronae. In French (315 × 190)

Fine border, initials. Penwork initials.

Contemporary binding of faded pink leather on boards.

Prov. Sir Lionell Tollemache of Helmingham, Suffolk (d. 1668), fols. i^{v}, 156^{v}, 157. Lancelot Lake, Lincoln's Inn 1633 (d. *c.* 1680), fol. ii.

386. **Balliol College 217** 1402
Tabula septem custodiarum, Part II (Comm. on the New Testament) (315 × 210)

Borders, initials (mutilated). '. . . tabula nova . . . incepta per manus Petri Wylhymleyd . . . 1401 . . . Completaque . . . 1402', fol. 116^{v}.

Prov. Cf. no. 503.

Mynors, *Catalogue*, xlii.

387. **University College 169** 1404
Barking
Ordinale and Customary of Barking Abbey (mutilated) (234 × 160)

Fine borders, initials. Fine penwork initials. [Illus.]

Prov. Commissioned and given to Barking abbey in 1404 by Sibilla de Felton, abbess 1394–1419, p. 2. Obits of Barking abbesses, fols. i–v, saec. xv. Obadiah Walker, master 1676–89. Sold at the sale of John Humphrey of Rothwell, 4 Dec., 1682, lot 20.

J. B. L. Tolhurst *ed*., 'The ordinale and customary of Barking Abbey', *H.B.S.* LXV–VI (1926–7); 'Lending books in a medieval nunnery', *B.Q.R.* v (1928), 188; Hunt, *Manuscript*, 26, 27 n. 1.

388. **Christ Church 148** After 1405 London (?)

John Gower, Confessio amantis (first version). In Latin and English (385 × 263)

Fine borders, initials. Written by the scribe of Bodl. MS. Bodley 294, Corpus Christi College 67 (no. 391) and 198 (no. 390), etc. (cf. Doyle, Parkes, 177).

Prov. Arms of England and France in the initial, fol. 1. Probably commissioned for Thomas, duke of Clarence, after 1405 (d. 1421). Edward Cottel, saec. xvi², first pastedown. Given by Henry Byam, B.D. (d. 1669), fol. 206ᵛ.

A. I. Doyle, M. B. Parkes, 'The production of copies of the *Canterbury Tales* and the *Confessio Amantis* in the early fifteenth century', *Medieval Scribes, Manuscripts and Libraries. Essays presented to N. R. Ker, eds.* M. B. Parkes, A. G. Watson, 1978, 177 no. 30, 195, Append. C 208.

389. **University Archives C.3** 1407–8 Oxford

Junior Proctor's Book (315 × 200)

Good border, initials (rubbed and damaged).

Prov. Copied at the expense of Richard Flemmyng, northern proctor of the University 1407–8, bp. of Lincoln 1419–31. His arms and the arms of the Waterton family in the initials, fols. 1, 9ᵛ.

S. Gibson, *Statuta antiqua Universitatis*, 1931, xiv–xv, 198–9; Emden, 697.

390. **Corpus Christi College 198** Saec. xv in. London (?)

Geoffrey Chaucer, The Canterbury Tales (mutilated). In English (322 × 218)

Fine borders, fine and other initials. Marginal instructions as to initials. Written by the scribe of Christ Church 148 (no. 388), Corpus Christi College 67 (no. 391), etc. [Illus.]

Prov. Given by William Fullman, fellow, rector of Meysey Hampton, Glos. (d. 1688), fol. 1.

J. M. Manly, E. Rickert, M. Rickert, *The Canterbury Tales*, i, 1940, 92 ff., 562–9, pl. ii; B. Boyd, *Chaucer and the medieval book* (The Huntington Library, San Marino), 1973, 48; A. I. Doyle, M. B. Parkes, 'The production of copies of the *Canterbury Tales* and the *Confessio Amantis* in the early fifteenth century', *Medieval Scribes, Manuscripts and Libraries. Essays presented to N. R. Ker, eds.* M. B. Parkes, A. G. Watson, 1978, 177, 178 n. 35, 195, 209, pl. 52.

391. **Corpus Christi College 67** Saec. xv in.

John Gower, Confessio amantis (first unrevised version). In English with Latin summaries (340 × 230)

Fine miniatures, borders, initials. Close in style to Herman Scheerre. Written by the scribe of Christ Church 148 (no. 388), Corpus Christi College 198 (no. 390), etc. [Illus.]

Prov. Thomas Crispe, London mercer, saec. xvi, fol. 209. Device containing names of the Crispe family, fol. 1ᵛ. Willelmus Rawson, 1580, fol. 1ᵛ.

G. C. Macaulay, 'The English works of John Gower', *E.E.T.S.* ex. ser. LXXXI (1900), cxlviii; G. M. Spriggs, 'Unnoticed Bodleian manuscripts illuminated by Herman Scheerre and his school', *B.L.R.* vii (1964), 198, pls. xii(b), xiii(a); J. H. Fisher, *John Gower*, 1965, 89, 117, 125, 352–4; A. I. Doyle, M. B. Parkes, 'The production of copies of the *Canterbury Tales* and the *Confessio Amantis* in the early fifteenth century', *Medieval Scribes, Manuscripts and Libraries. Essays presented to N. R. Ker, eds.* M. B. Parkes, A. G. Watson, 1978, 177, 178 n. 35, 195, 209, pl. 50; K. L. Scott, 'John Lydgate's Lives of Sts Edmund and Fremund', *Viator*, xiii (1982), 337 n. 8.

392. **All Souls College 98** Saec. xv in.

John Gower, Vox clamantis (revised version) and other poems (302 × 210)

Fine initials. Good penwork initials. Preceded by a dedicatory poem, fol. 1ᵛ, a nearly contemporary addition with a historiated initial close in style to Herman Scheerre. Partly written by the scribe of Glasgow, Hunterian Museum MS.T.2.17. [Illus.]

Prov. Thomas Arundel, archbp. of Canterbury 1396–1414, to whom the poem is addressed, represented in the initial on fol. 1ᵛ.

G. C. Macaulay, *ed*., *The complete works of John Gower*, v. iv, 1902, lix–lxii; J. H. Fisher, *John Gower*, 1965, 89, 99–101, 112–13, 117, 125–6, 346, 350, 352–3, 355; A. I. Doyle, M. B. Parkes, 'The production of copies of the *Canterbury Tales* and the *Confessio Amantis* in the early fifteenth century', *Medieval Scribes, Manuscripts and Libraries. Essays presented to N. R. Ker, eds.* M. B. Parkes, A. G. Watson, 1978, 163–4 n. 3.

393. **St. John's College 179** Saec. xv in.

Breviary. Use of Sarum. Pars hiemalis (190 × 122)

Fine borders, initials. Good penwork initials.

Prov. Stephen Pott, London bookseller, 1610, fol. 1.

394. **Corpus Christi College 12,** fols. 2–132, 210–58 Saec. xv in.
Bible (Genesis-Kings, Psalter) (370 × 258)

Good and other initials. Penwork initials.

Prov. Richard Watkins, scholar of Brasenose College, *c.* 1705/6, fol. 93.

395. **University College 56** Saec. xv in.
Richard Rolle, Comm. on the Psalter (297 × 190)

Fine penwork borders, initials.

Prov. Calendar of Christ Church, Canterbury, with obits of the Hampton family, 1413–22, fols, 1–5^{v}. Given by Samuel Clarke, commoner 1641, fols. 1, 264^{v}.

Hunt, *Manuscript*, 18, 20.

396. **Queen's College 324** Saec. xv in.
Primer. In English (235 × 150)

Fine borders. Fine and other initials. Good penwork initials.

Medieval binding of brown leather on boards.

397. **Queen's College 388** Saec. xv in.
Bible. In English (earlier Wycliffite version; incomplete) (420 × 298)

Good and other initials (mutilated). Penwork initials.

Prov. William Smyth of Scultcham, farmer 1561–3, fol. 8^{v}.

398. **All Souls College 92** Saec. xv in.
Aegidius de Columna, de regimine principum (282 × 190)

Good penwork borders, initials.

Prov. Given before 1495 by Thomas Lay, fellow, fol. 1^{v}.

N. R. Ker, *Records of All Souls College Library 1437–1600*, 1971, 124.

399. **Merton College 117** (L.3.4) Saec. xv in.
Robert Cowton, Comm. on the Sentences (376 × 252)

Fine borders, initials. Good penwork initials.

400. **Oriel College 75** Saec. xv in.
Missal. Use of Sarum (313 × 208)

Fine miniature, borders in two styles. Historiated and other initials. The miniature and historiated initials close in style to Johannes and his school (cf. Bodl. MS. Bodley 264). Fine marginal drawings (catchwords, explicits), sketches, fols. 32^{v}, 44^{v}. Written by Sampson, fols. 211, 319. [Illus.]

Prov. High grading of St. Oswald (calendar fols. 1–6^{v}) suggests the diocese of Worcester but the Feast of relics is that of Salisbury Cathedral.

J. Wickham Legg, 'Tracts on the Mass', *H.B.S.* XXVII (1904), xiv.

401. **Exeter College 41** Saec. xv in.
Johannes de Burgo, Pupilla oculi. Religious miscellanea (325 × 225)

Fine borders, initials. Penwork initials.

402. **Exeter College 7,** fols. 6–163 Saec. xv in.
William Wodeforde, de sacramento Eucharistie and other texts (217 × 137)

Fine border, initial. Penwork initials. 'Chyn' (scribe ?), fol. 163.

403. **Corpus Christi College 350** Saec. xv in.
Book of Hours. Use of Sarum (220 × 123)

Fine borders, historiated and other initials. By the workshop of Herman Scheerre (cf. nos. 391, 392). [Illus.]

Prov. Made for a lady kneeling before the Virgin, fol. 7. Notes relating to the Wharton family, saec. XVI, fol. iii^{rv}. Given by John Harrison, fellow, B.D. 1727, fol. 1.

G. M. Spriggs, 'Unnoticed Bodleian Manuscripts by Herman Scheerre and his school', *B.L.R.* VII (1964), 201 n. 2; K. L. Scott, 'John Lydgate's Lives of Sts Edmund and Fremund', *Viator*, XIII (1982), 347–9 n. 50.

404. **Corpus Christi College 193** Saec. xv in.
Richard Rolle, Comm. on the Psalter, etc. (270 × 185)

Borders, good and other initials. 14 leaves, fols. 181–94, now bound separately as Brit. Libr. MS. Cotton Tib. A. xv.

Contemporary binding of white leather on boards.

Prov. Arms of Robert de Lacy, founder of Pontefract priory, Yorks. (1090), fol. 2. Fr. John Hanton, monk at St. Mary's, York, first pastedown, saec. xv in. Pressmark of St. Mary's, 'In viii G', fol. 1^{v}, saec. xv. H. Savile (d. 1617).

A. G. Watson, *The manuscripts of Henry Savile of Banke*, 1969, 29 (56).

405. **Corpus Christi College 161** Saec. xv in.
Speculum humanae salvationis (297 × 195)

Fine outline drawings with coloured washes. Fine border, initials, penwork initials. [Illus.]

Prov. Given by Cuthbert Nicholl to his grandson, John Clem, 1612, fol. 1^{v}.

A. Gaily, A. Fenton, *eds.*, *The Spade*, 1970, 30, fig. a(f); A. Heimann, 'Die Hochzeit von Adam und Eva in Paradise', *Wallraf Richartz Jahrbuch*, XXXVII (1975), 18, Abb. 7.

406. **Christ Church 87** Saec. XV in.
Missal. Use of Sarum (425 × 273)

Fine penwork initials.

Contemporary binding of white leather on boards.

Prov. Written for St. Botolph's church, Aldgate, London (dedication of the church in the calendar).

J. Wickham Legg, 'Tracts on the Mass', *H.B.S.* XXII (1904), xiv.

407. **Corpus Christi College 4** Saec. XV in.
Bible. In English (earlier Wycliffite version) (475 × 290)

Penwork initials.

Prov. Given by A. Langford, secretary to Thomas Bilson, bp. of Winchester, 1615, fol. 1.

Forshall and Madden, no. 94; S. L. Fristedt, *The Wycliffe Bible*, pt. I, 1953, 15.

408. **Balliol College 93** Saec. XV in.
Walter Burley, de potenciis anime, etc. (incomplete) (350 × 217)

Good initials (mutilated). Penwork initials. Good marginal drawings, especially fols. 45^v, 85^v, 131^v, saec. XV. [Illus.]

Prov. William Gray, bp. of Ely 1454–78, fol. 2^v.

409. **University College 62–3,** fols. 1–64 Saec. XV in.
Robert Grosseteste, de lingua et corde (255 × 175)

Good border, initial.

Binding of blind stamped leather on boards, marks of a label on the lower cover, saec. XVI in., possibly A. Lysley, Eton.

Prov. Humphrey Gilbert, *c.* 1551, fol. 64^v and on lower cover. Given by Maria Elmhurst in memory of Herbert Elmhurst, scholar (d. 1670), first pastedown.

Hunt, *Manuscript*, 20; N. R. Ker, 'The Virgin and Child Binder', *The Library*, XVII (1962), 79.

410. **Keble College 58** Saec. XV in.
Missal. Use of Sarum (318 × 208)

Fine borders, initials. Good penwork initials.

Prov. William Bragge sale, Sotheby, Wilkinson and Hodge's, 7 June 1876, lot 468. Sir Thomas Brooke, Huddersfield (d. 1908). Bequeathed by his brother, Canon C. E. Brooke (d. 1911).

411. **Magdalen College Lat. 33** Saec. XV in.
Sermones dominicales (252 × 172)

Fine border, initial. Penwork initials.

Prov. 'Precium huius libri xxxij s.', fol. 192.

412. **Magdalen College Lat. 47** Saec. XV in.
Jo. Tarteys, Problemata in Porphyri Isagogen. Wm. Mylverley, Universalia. W. Burley, Comm. on Porphyri Isogogen, etc. (220 × 154)

Penwork initials. Drawing (diagram), fol. 54.

Prov. Request for prayers for Thomas Loughburgh, canon of Lincoln (d. 1450) and William Kirkgarth, rector of St. Mary's, Lincoln, *c.* 1462, fol. 2.

Emden, 1079, 1164.

413. **Merton College 251** (C.1.13) Saec. XV in.
John Sharpe, Quaestiones de anima. Miscellanea (190 × 135)

Good borders, initials. Penwork initials. Written by Wyke, fol. 157. Part of Merton 251 is now Bodl. Digby 77, fols. 150–97.

Prov. Probably given in 1425 by John Whethamstede, abbot of St. Albans 1420–40, to Gloucester College, Oxford, fol. 74.

Powicke, *Medieval Books*, no. 1202.

414. **Balliol College 192** Saec. XV in.
Duns Scotus, Quodlibeta abbreviata, etc. (297 × 199)

Initial. Penwork initials.

Prov. Notes by Thomas Gascoigne, chancellor of Oxford University (d. 1458), fols. 1, 2, 2^v, 63.

415. **Trinity College 23** *c.* 1420
Gospel lectionary. Use of Sarum (270 × 195)

Fine borders. Fine and other initials. Good penwork initials. [Illus.]

Prov. Given in 1428 by Henry Chichele, archbp. of Canterbury 1414–43, to Higham Ferrers College, Northants., its name erased and replaced by that of All Souls College, Oxford, saec. XVI (?), fol. 10.

Ker, *Medieval Libraries*, 101 n. 1, 144, 269.

416. **Magdalen College Lat. 97** *c.* 1420
Ranulf Higden, Polychronicon, with continuation to 1377 (325 × 225)

Fine borders, initials. Good marginal drawings (catchwords). [Illus.]

Original binding of white leather on boards.

Prov. Alexander Hill, 1572, fol. 207^v. Given by Francis Hill, fellow, clerk of the College 1628–30.

J. W. Taylor, 'The development of the Polychronicon continuation', *English Historical Review*, LXXVI (1961), 27, 35; *id.*, *The Universal Chronicle of Ranulf Higden*, 1965, 116, 157, 178.

417. **Lincoln College Lat. 106** *c.* 1420
Thomas Netter of Walden, Doctrinale fidei (420 × 280)

Fine historiated initials, borders. Fine and other penwork initials. [Illus.]

Prov. English royal arms added, fol. 3. Thomas Gascoigne, chancellor of Oxford University (d. 1458), first pastedown. Given by the founder, Richard Flemmyng, bp. of Lincoln (d. 1431), fol. 1^{v}.

Weiss, *Catalogues*, 348 (25) n. 48; Emden, 698, 746; Weiss, *Humanism*, 61.

418. **St. John's College 94** *c.* 1420
Newcastle-upon-Tyne
Miscellaneous devotions. Books of Hours. Use of Sarum. In Latin and English. (265 × 175)

Good miniatures (mutilated and partly defaced), that of fol. 16^{v} dated 1420. Fine borders, fine and other initials. Penwork borders, initials, line-fillers. Written and illuminated by John Lacy, Dominican recluse, fols. 16^{v}, 17. [Illus.]

Prov. Arms of John Lacy, fol. 101^{v}, who gave the MS. in 1434 to Roger Stonysdale, chaplain, for the use of St. Nicholas', Newcastle-upon-Tyne, fol. 1. Henry Savile (d. 1617).

R. M. Clay, 'Further Studies on Medieval Recluses', *Journal of the British Archeological Association*, 3rd ser., XVI (1953), 75–8, pl. XXIIA; A. G. Watson, *The Manuscripts of Henry Savile of Banke*, 1969, no. 60.

419. **University College 168** Saec. xv$\frac{1}{4}$
Collection of commentaries on the Song of Songs, etc. (384 × 250)

Good historiated initial. Fine borders in an unusual style, fine and other initials. Fine penwork initials. Marginal drawings (catchwords), later pencil sketches, saec. XVI.

Contemporary binding of white leather on boards.

420. **New College 66** Saec. xv$\frac{1}{4}$
Old Testament. In English (later Wycliffite version) (305 × 212)

Fine borders, initials. Good penwork initials.

Forshall and Madden, no. 97; S. L. Fristedlt, *The Wycliffe Bible*, pt. I, 1953, 104.

421. **Balliol College 50** Saec. xv$\frac{1}{4}$
Hugo Ripelin, Compendium totius theologie, etc. (292 × 125)

Initials. Penwork initials.

Prov. Robert Pollan, fellow, vicar of Penistone, Yorks. (d. 1459), fols. iii^{v}, 110^{v}.

Emden, 1525.

422. **University College 78A** Saec. xv$\frac{1}{4}$
Hereford
Missal. Use of Hereford (mutilated) (287 × 191)

Fine borders, fine and other initials. [Illus.]

Original boards, rebacked.

Prov. Dedication of Hereford Cathedral in the calendar. Dedication of St. Dubricius, Whitchurch, Monmouthshire, probably a contemporary addition. Given in 1668 by William Rogers of Painswyck, commoner, fols. 1, 7, 227^{v}.

Hunt, *Manuscript*, 21.

423. **University College 37** Saec. xv$\frac{1}{4}$
Catena SS. Patrum in psalmos septem penitentiales (210 × 140)

Fine borders, initials.

424. **Balliol College 263,** fols. 7^{v}–33^{v} Saec. xv$\frac{1}{4}$
Galfridus de Vino Salvo, de arte versificandi, etc. (250 × 180)

Fine borders (damaged by damp), initials.

425. **Magdalen College Lat. 30** Saec. xv$\frac{1}{4}$
Vegetius, de re militari. English translation by John Walton made in 1408 (247 × 152)

Fine borders, initials. [Illus.]

Prov. Translated for Thomas Berkeley (d. 1417), fol. 124^{v}, his arms on fol. 2.

A. J. Perry, 'Notes on John Trevisa', *Modern Language Notes*, XXXIII (1918), 15, 16; M. Science, 'Boethius, *De consolatione philosophiae* translated by John Walton, canon of Osney', *E.E.T.S.* or. ser. CLXX (1927), xlviii, xlix.

426. **Exeter College 37** Saec. xv$\frac{1}{4}$
Augustine, Comm. on Psalms CI–CL (305 × 217)

Good border, initial. Penwork initials.

427. **Oriel College 20** Saec. xv$\frac{1}{4}$
Robert Grosseteste, Dicta, etc. (283 × 201)

Fine borders (initials mutilated). Good penwork initials.

Prov. Given by Robert Grafton, fellow 1456 (still in 1466–7), fols. 1, 310^{v}.

Emden, 799.

428. **Wadham College 5** (A.5.8) Saec. xv$\frac{1}{4}$
Ps.-Bonaventura, The mirror of the life of Christ. English translation by Nicholas Love (240 × 165)

Borders (rubbed), initials. Penwork initials.

Prov. Given by Philip Stubbs, fellow, in 1695.

429. **St. John's College 109** Saec. xv$\frac{1}{4}$
John Felton, de pera peregrini (247 × 155)

Border, initial. Penwork initials.

Prov. Given to Osney abbey by John Holbech, prior, who bought it for xiiij s., iiij d., fol. 245. Pressmark 'M', fol. 1, saec. xv^2. Richard Lateware, fellow, 1598, fol. 1.

430. **Brasenose College 9** Saec. xv$\frac{1}{4}$
Ps.-Bonaventura, The mirror of the life of Christ. English translation by Nicholas Love (240 × 187)

Good borders, initials. Penwork initials.

Prov. Initials 'N.L.', fol. 5. Possibly a copy made for the translator, Nicholas Love, prior of Mount Grace, Yorks., 1410.

Meditations on the life of Christ, eds. I. Ragusa, R. Green, 1961, xxiii n. 5.

431. **Queen's College 304** Saec. xv$\frac{1}{4}$
Glastonbury (?)
Peter of Ickham, John Merylynch, Chronica Angliae, etc. (370 × 245)

Fine borders, historiated and other initials. Good penwork initials. Genealogical tables. Preliminary sketch, fol. 144^v. 'Per me Gerardus Bacaveridis' (scribe ?), fol. 46^v. [Illus.]

Prov. '. . . de perquisito . . .' of John Merylynch, monk of Glastonbury (cf. Bodl. MS. Laud Lat. 4, *c.* 1406). Arms of Glastonbury abbey, fol. 1. Arms of Keyes of Kent, fol. 50.

432. **St. John's College 167** *c.* 1419–26
Processional (191 × 136)

Good initial. Woodcut added in the margin, fol. 5, saec. xv. Penwork initials.

Prov. Written for the use of the Bridgettine Sisters of St. Saviour, Twickenham (from 1431 at Syon, Middlesex), fol. 100^v. Datable on liturgical grounds (cf. Pfaff). Sister Mary Nevel (d. 1558), fol. 1.

R. W. Pfaff, *New Liturgical Feasts in Later Medieval England*, 1970, 29 n. 1, 57 n. 1.

433. **Keble College 66** *c.* 1420–30
Book of Hours (fragment). Use of Sarum (197 × 147)

Fine borders, initials. [Illus.]

Prov. Bequeathed by Canon C. E. Brooke (d. 1911).

434. **All Souls College 85** 1427–8
John Sharpe, Comm. on Aristotle, Physics, etc. (290 × 185)

Penwork border, initial. Written by Albertus 'nacione Brabancie', fols. 87^v, 133^v, 276^v.

Prov. Arms of Winchcomb abbey, Glos., in the initial, fol. 1.

435. **Oriel College 70** 1429
Oxford
Scotulus, Quaestiones on the Sentences (275 × 185)

Good penwork borders, initials. Written by Nicolaus de Bodelswerdia, an Oxford scholar, fol. 168^v, who also wrote MSS. Merton College 133 in 1427 and Lincoln College Lat. 93.

Original cover and pastedowns preserved as fly-leaves.

Exh., *Duke Humfrey*, nos. 67–8.

436. **Merton College 297B** (K.1.7) Before 1435
Index of Statutes. Misc. texts. Statutes of England from Edward III to 1435 with continuation to 23 Henry VI. In Latin and French (290 × 200)

Fine border, historiated (one) and other initials. Good border, historiated initials in a different style and probably later, saec. xv med. (?). Marginal sketch, fol. 275^v. Good penwork initials, those on fols. 340^v–91 presumably added at Canterbury, 1435–44. [Illus.]

Prov. Calendar of St. Paul's, London (?). Bequeathed to Christ Church, Canterbury, in 1435 by John Pyrye, a London jurist, fols. 9^v, 110^v, 214^v, 299^v. Given by James Leach, fellow (d. 1589), fols. i, ii.

Powicke, *Medieval Books*, 1255.

437. **Magdalen College Lat. 39** *c.* 1437
Humanistic miscellanea (276 × 208)

Fine initials imitating Italian white-vine decoration. By the same hand as Bodl. MS. Bodley 915 and additions to New College, Oxford MS. 271 (no. 609), possibly by Thomas Candour who wrote the MS. in humanistic script on Italian parchment probably *c.* 1437 and later. Cf. no. 575. For other MSS. see Index.

Prov. Samuel Wrott, armiger, tenant of the College, fol. ii, saec. XVI.

Exh., *Duke Humfrey*, no. 57 pl. x; Pächt, Alexander, 3, no. 1088; N. Mann, 'La prima fortuna del Petrarca in Inghilterra', *Il Petrarca ad Arqua*, 1975, 285 n. 3; *id.*, 'Petrarch manuscripts in the British Isles', *Italia medioevale e umanistica*, XVIII (1975), no. 241; A. C. de la Mare, 'Humanistic hands in England', Exh. *Manuscripts at Oxford*, XXII, 2.

438. **Balliol College 122** *c.* 1440
Oxford (?)
Valerius Maximus, Memorabilia, with comm. of Dionysius of Borgo San Sepolcro (370 × 280)

Good border, initials (mutilated). The commentary written in humanistic script by the scribe of Balliol 276 (no. 442).

Prov. William Gray, bp. of Ely 1454–78, fol. 2ᵛ.

Exh., *Duke Humfrey*, 25–6.

439. **Christ Church 145** *c.* 1440
Bible. In English (earlier Wycliffite version) (390 × 255)

Good border, initial. Penwork initials.

Contemporary binding of white leather on boards.

Prov. Given by Robert Clay in 1575 to Edward Saunders, M.A. 1586, who gave it to the College, fol. 1.

Forshall and Madden, no. 91; S. L. Fristedt, *The Wycliffe Bible*, pt. I, 1953, pl. XIII.

440. **Christ Church 146** *c.* 1440
New Testament. In English (later Wycliffite version) (195 × 130)

Initials. Penwork initials.

Binding of stamped leather on boards.

Prov. Thomas Ballowe, M.A., 1629.

Forshall and Madden, no. 92.

441. **Lincoln College Lat. 101** *c.* 1440
Thomas Waleys, de theoria predicandi. Ps.-Dionysius, de divinis nominibus, etc. (330 × 207)

Fine initials, borders. Good penwork initials. Written by John Malberthorpe, fellow 1436, who gave it to the College.

Binding of white leather (damaged), on original boards.

Prov. 'Laus tibi Jhesu semper dulcedo mi. M.J.M.', motto of the scribe, fols. 47ᵛ, 95.

Weiss, *Catalogues*, 349 (54) n. 78; N. R. Ker, 'Chaining from a staple on the back cover', *B.L.R.* III (1950–1), 104; G. Pollard, 'Describing Medieval Bookbindings', *Medieval Learning and Literature. Essays presented to R. W. Hunt, eds.* J. J. G. Alexander, M. T. Gibson, 1976, 61.

442. **Balliol College 276,** fols. 2–56ᵛ *c.* 1442
Oxford

Galfridus de Vino Salvo, Poetria nova, etc. (incomplete) (340 × 255)

Good border, initials (mutilated). Good penwork border, initials. Humanistic script by the scribe who also wrote for William Gray MSS. Balliol 122, 126, 315 (nos. 438, 935, 444). Cf. no. 850.

Prov. William Gray, bp. of Ely 1454–78, fol. 1ᵛ.

Mynors, *Catalogue*, xxvii; Exh., *Duke Humfrey*, nos. 40, 41.

443. **Balliol College 28** 1442
Oxford (?)

Thomas Docking, Comm. on Deuteronomy (397 × 265)

Good border, initial. Written for William Gray by a Dutch scribe, Tielman Reynerszoon, probably in Oxford, in 1442, colophon fol. 258. Cf. nos. 511, 928.

Prov. William Gray, bp. of Ely 1454–78, fol. 4ᵛ. His arms in the initial fol. 5. Notes by Thomas Gascoigne, chancellor of Oxford University (d. 1458).

Mynors, *Catalogue*, xxvi; Ker, *College Libraries*, 309 n. 81.

444. **Balliol College 315** *c.* 1442
Oxford (?)

Humanistic miscellanea incl. Pliny, Panegyricon, etc., fols. 69–148 (260 × 180)

Good borders, initials. Penwork initials. Close in style to MS. Balliol 122. Written for William Gray in humanistic script probably in Oxford *c.* 1442 by the scribe of MSS. Balliol 122, 126, 276 (nos. 438, 935, 442). The Panegyricon is a direct copy of Duke Humfrey's MS., now Paris, Bibl. Nat. Latin 7805.

Prov. William Gray, bp. of Ely 1454–78, fol. 3ᵛ.

Mynors, *Catalogue*, xxvii–viii; Exh., *Duke Humfrey*, 6, no. 40, pl. v; Ker, *College Libraries*, 305.

445. **Oriel College 32** Before 1444
John Capgrave, Comm. on Genesis (412 × 265)

Fine historiated initials, borders, diagram. Good penwork initials. Companion volume with Bodl. MS. Duke Humfrey b. 1 (which is illuminated by a different hand). Possibly written in King's Lynn Augustinian priory, Norfolk, supervised and corrected by John Capgrave. [Illus.]

Remains of the original leather cover and paste-downs used as fly-leaves.

Prov. Presentation copy to Humfrey, duke of Gloucester, 1439, his *ex libris* on fol. 187ᵛ. Given by him to the Old University Library, Oxford, 1444, fol. 3.

A. de Meijer, 'John Capgrave O.E.S.A.', *Augustiniana*, VII (1957), 134, 145, 531, 545, pl. opp. p. 552; Ker, *Medieval Libraries*, 143, 289; P. J. Lucas, 'John Capgrave O.S.A. (1393–1464), scribe and publisher', *Transactions of the Cambridge Bibliographical Society*, V (1969), 1 ff., pls. VI, VII; Pächt, Alexander, 3, no. 902; E. Colledge, 'The Capgrave Autographs', *Transactions of the Cambridge Bibliographical Society*, VI (1974), 137 ff.

446. **Magdalen College Lat. 156** 1448
Homiliary (415 × 280)

Good borders (mutilated), initials. Written by Thomas Colyngborne who also wrote MSS. Magdalen College 145 (no. 458) and 154.

Original binding of white leather on boards.

Prov. Given jointly by John Dygoun, recluse of Sheen (adm. 1435), and Johanna Grenewode, reclusa of St. Botulph's, Bishopgate, London, to Thomas Grenewode with the reversion to the College, fol. iv.

R. M. Clay, 'Further studies in medieval recluses', *Journal of the British Archaeological Association*, 3rd ser., XVI (1953), 78 n. 7; Emden, 616; M. B. Parkes, *English cursive book hands 1250–1500*, 1969, pl. 24 (i).

447. **Balliol College 310** *c.* 1449
Leonardo Bruni, Letters (292 × 200)

Fine borders, initials. Penwork initials. Humanistic script by Theodericus Werken, 1449, colophon fol. 153. For other MSS. see Index.

Prov. Richard Bole, fellow, archdeacon of Ely (d. 1477), fol. 1v.

Mynors, *Werken*, 101, 104; Exh., *Duke Humfrey*, no. 51; de la Mare, *A fragment*, 285 ns. 3, 9.

448. **Worcester College 233** Saec. xv $\frac{2}{4}$
St. Albans
Thomas Netter of Walden, Doctrinale fidei, etc. (385 × 275)

Fine historiated penwork borders, fine penwork initials. By the same hand as Bodl. MS. Auct. F. infra 1. 1, fol. 169 and New College, Oxford, MS. 49 (cf. no. 449). Written in large part by the scribe of the latter.

Binding of white leather on original boards, repaired, saec. XV.

Prov. Vol. II of a set (vol. I is now Brit. Libr. MS. Royal 8. C. X) made for John Whethamstede, abbot of St. Albans, who gave it to Gloucester College, Oxford, *c.* 1450. Given to Merton College in 1558 (MS. 318) by Robert Serles, fellow, vicar of St. Peter's-in-the-East, Oxford. Returned to Worcester College in 1938.

Powicke, *Medieval Books*, no. 1201; D. R. Howlett, 'Fifteenth-century manuscripts of St. Albans Abbey and Gloucester College Oxford', Exh. *Manuscripts at Oxford*, XX, 3; D. R. Howlett, R. W. Hunt, 'The provenance of MS. New College 49', *B.L.R.* X (1981), 225–8, pls. XII, XV.

449. **New College 49** Saec. xv $\frac{2}{4}$
St. Albans
Petrus Johannes Olivi, Comm. on St. Matthew (430 × 285)

Fine historiated penwork border, initials. Same iconography and style as Bodl. MS. Auct. F. infra I. 1, fol. 169, and Worcester College MS. 233, fols. 2, 84 (cf. no. 448), made *c.* 1420–30 for John Whethamstede, abbot of St. Albans. The scribe also wrote a large part of the latter.

Contemporary binding of white leather on boards with a staple on the back cover.

Prov. Probably given by John Whethamstede to Gloucester College, Oxford.

N. R. Ker, 'Chaining from a staple on the back cover', *B.L.R.* III (1950–1), 105–7, figs, 2, 3; Hunt, *Medieval Library*, 318, n. 3; D. R. Howlett, 'Fifteenth-century manuscripts of St. Albans Abbey and Gloucester College Oxford', Exh., *Manuscripts at Oxford*, XX, 4; D. R. Howlett, R. W. Hunt, 'The provenance of MS. New College 49', *B.L.R.* X (1981), 225–8, pl. XIV.

450. **St. John's College 172** Saec. xv $\frac{2}{4}$
Oxford (?)
Richard de Bury, Philobiblon, etc. (199 × 130)

Initials.

Prov. Ex libris of Oxford University saec. XV, first pastedown. John Alword, fellow of Exeter College, rector of Stoke Bruere (d. 1458), fol. 2. Sir William Paddy, commoner, physician to James I (d. 1634), fol. 2.

451. **Exeter College 36,** fols. 1rv, 210–35v Saec. xv $\frac{2}{4}$
Bible (335 × 227)

Fine borders, initials added to a MS. of saec. XIII in.

Prov. Deposited jointly in the Winton chest, Oxford, by magister John Mercer, fellow of Magdalen College, and magister William Masse, 3 Nov. 1479, fol. 4v.

Emden, 1241, 1262.

452. **Corpus Christi College 162** Saec. xv $\frac{2}{4}$
Johannes de Burgo, Pupilla oculi (300 × 200)

Fine border, penwork initials.

Prov. Given by Henry Parry, fellow, 1619, fol. 1.

453. **St. John's College 47** Saec. xv $\frac{2}{4}$
Manual. Use of Sarum. In Latin and English (310 × 205)

Good borders, good and other initials. Penwork initials.

Prov. Richard Butler, archdeacon of Northampton (d. 1612). Given in 1613 through the good offices of John Buckeridge, bp. of Rochester and Ely, fol. 2.

A. S. Collins *ed.*, 'Manuale ad usum percelebris ecclesie Sariburiensis', *H.B.S.*, XCI (1958), 56 n. 68, 98 n. 3, 123 n. 38, 140 n. 52.

454. **Balliol College 186** Saec. xv $\frac{2}{4}$
Albertus Magnus, Comm. on St. Matthew's Gospel (350 × 245)

Border, initials (mutilated).

Prov. Robert Abdy, fellow (d. 1483), fol. 1v.

Emden, 2; N. R. Ker, 'Eton College MS. 44 and its

Exemplar', *Litterae textuales. Essays presented to G. I. Lieftinck* I, *eds.* J. P. Gumbert, M. J. M. De Hann, 1972, 59–60.

455. **Lincoln College Lat. 112** Saec. xv^2_4
Ralph Acton, Sermons (375 × 254)

Good border (mutilated), initial. Penwork initials. 'Sutton', fol. 144, possibly the scribe.

Medieval binding of white leather on boards, saec. xv.

456. **Lincoln College Lat. 151** Saec xv^2_4
Brut Chronicle (incl. the reign of Henry V). In English (240 × 172)

Good border, initial (defaced). Penwork initial, marginal scribbles.

Prov. Given by George Norris to Norwich Castle Museum, 1930. Sotheby's sale, 11 December 1961, lot 159. Given by Mr. Eric Sexton, Camden, Maine, U.S.A., *c.* 1972–3, first pastedown.

457. **Keble College 13** Saec. xv^2_4
Book of Hours, Use of Sarum. In Latin and English (149 × 106)

Fine borders. Fine and other initials, penwork initials. Added leaves, fols. 167–80^v, written in the Netherlands in saec. xv^1 (rubric in Dutch), were illuminated in England by a different hand.

Prov. Sir Hugh Patersone, bart, 1709, first pastedown. Bequeathed by Dr. Henry P. Liddon, canon of St. Paul's, London (d. 1890).

458. **Magdalen College Lat. 145** Saec. xv^2_4
Nicolas de Byard, Distinctiones (420 × 280)

Fine borders, initials. Marginal drawings. Lombards. Written by Thomas Colynborne who also wrote MSS. Magdalen College 154 and 156 (no. 446) in 1448.

Binding of stamped leather on boards, the same stamps used on the binding of Bodl. MS. Bodley 117 written by William Mede, monk of the Charterhouse, Sheen, saec. xv^2.

Prov. Notes by John Dygoun, recluse of Sheen, Surrey (adm. 1435).

459. **Lincoln College Lat. 52** Saec. xv^2_4
Fasciculus morum (300 × 190)

Good border, initial. Good penwork initials.

Prov. Robert Flemmyng, dean of Lincoln (d. 1483), fol. i^v.

Emden, 700.

460. **Balliol College 219** Saec. xv^2_4 Oxford

Alphabetum narrationum, etc. (181 × 110)

Initials (mutilated). Penwork initials. Marginal drawings, fols. 157^v, 168^v. Written for and largely by Richard Stapleton, master of the College, who also wrote part of Bodl. MS. Digby 29, *c.* 1430 (cf. Pächt, Alexander, 3, no. 889).

Prov. Richard Stapleton, fol. 238^v.

Emden, 1766.

461. **Christ Church 121** Saec. xv^2_4
Gesta Romanorum (210 × 145)

Initial. Penwork initials.

Contemporary binding of white leather on boards.

Prov. Matthew Weys, 'alumnus et theologus', 1640, fol. 1.

462. **All Souls College 10** Saec. xv^2_4
Gospels with gloss. French translation by Guyart Desmoulins (292 × 200)

Fine borders, historiated and other initials, line-fillers. Good penwork initials. 'Deo gracias. Doal', possibly the scribe, fol. 204^v. Close in style to Karlsruhe, Badische Landesbibl. S. Georgen 12, and the Clarence Hours, sold Sotheby's, 12 December 1967, lot 46. [Illus.]

Medieval binding of white leather on boards.

Prov. William Kingsley, fellow, prebendary of Canterbury cathedral, 1614, fol. 3.

S. Berger, *La Bible française au moyen age*, 1884, 412.

463. **Magdalen College Lat. 189** Saec. xv^2_4
Aristotle, Ethics and Politics, with gloss (310 × 200)

Good initials. Penwork initials.

Contemporary binding of blind-stamped leather on boards. Originally four bosses on each cover, now missing. Oxford *c.* 1470.

Prov. Nicholas Good, fellow (d. 1479). Given in his memory in 1502–3 by William Attwater, fellow, fol. i^v.

S. Gibson, *Early Oxford Bindings*, 1903, 16, pl. IV; Emden, 789.

464. **Magdalen College Lat. 143** Saec. xv^2_4
William Lynwood, Provinciale, etc. (410 × 290)

Good penwork initials.

Medieval binding of white leather on boards, clasp.

Prov. William de Laguna (d. 1475), fol. 308^v.

Emden, 1084.

465. **Queen's College 210** Saec. xv$\frac{2}{4}$
Prayers of St. Brendan. In English. Book of Hours. Use of Sarum (202 × 137)

Good borders, initials. Cf. no. 830.

Prov. Given by Hugh Todd, fellow of University College (d. 1728), fol. i.

466. **Magdalen College Lat. 153** Saec. xv$\frac{2}{4}$
Thomas Netter of Walden, Doctrinale fidei (425 × 250)

Fine borders, historiated and other initials. Penwork initials. Written by fr. Richard Wetyng, a Carmelite, fol. 319^{v}. [Illus.]

Prov. '1594 . . . precium huius libri 6 li, 13 s. 4 d.', first pastedown. Given by Bartholomeus Jessope, D. C. L. Magdalen College, 1599 (d. 1620), fol. i.

467. **Magdalen College Lat. 160** Saec. xv$\frac{2}{4}$
William of Nottingham, Comm. on the Gospels (460 × 280)

Fine borders, initials. Good penwork initials.

B. Smalley, 'Which William of Nottingham', *Mediaeval and Renaissance Studies*, III (1954), 217.

468. **Pembroke College 1** Saec. xv$\frac{2}{4}$
Processional (incomplete). Use of Sarum (425 × 290)

Fine borders, fine and other initials (mutilated).

Original boards with white leather cover, straps (missing).

469. **New College 319** Saec. xv$\frac{2}{4}$
Boethius, de consolatione philosophiae. English translation by John Walton (270 × 195)

Good and other initials (partly defaced). Marginal drawing, fol. 103.

M. Science, 'Boethius, *De consolatione philosophiae* translated by John Walton, canon of Osney', *E.E.T.S.* or. ser. CLXX (1927), xiv, xlix.

470. **New College 115** Saec. xv$\frac{2}{4}$
(1) Scotulus, Quaestiones. (2) Opuscula commutationis (317 × 220)

Fine borders, historiated (one) and other initials. Penwork initials. [Illus.]

Prov. Thomas Hylle, fellow, rector of St. Ebbe's, Oxford (d. 1469), possibly the author of (2), dedicated to William Westbury, fellow, headmaster of Eton College, 1443, fol. 185.

Emden, 935, 2021.

471. **New College 87** Saec. xv$\frac{2}{4}$
Philip de Janua, Postillae (410 × 280)

Fine and other initials.

472. **Merton College 319** (Q.3.1) Saec. xv$\frac{2}{4}$
Thomas Netter of Walden, Doctrinale fidei (465 × 310)

Fine historiated and other initials (mutilated). Good penwork initials. Written by John Holt, a Carmelite friar, fols. 193, 323^{v}. [Illus.]

Prov. Bequeathed by John Warkworth, fellow, master of Peterhouse, Cambridge (1473–1500).

Powicke, *Medieval Books*, no. 1140.

473. **Trinity College s.n.** Saec. xv$\frac{2}{4}$
Missal. Use of Sarum. (Thomas Pope's Missal) (271 × 183)

Good borders, initials (mutilated). Penwork initials.

Prov. Sir Thomas Pope, the founder (d. 1559), fol. 270. William Braban, 1774. Sir R. Leicester Harmsworth, Bt., LL.D. (d. 1945). Bought Sotheby's sale, 16 October 1945, lot 2030.

474. **Magdalen College Lat. 218** Saec. xv$\frac{2}{4}$
Sallust, Bellum Catilinarium. Cicero, de officiis, etc. (179 × 125)

Good initial (damaged). Penwork initials. Written in part by Johannes Rybow, Anglicus, fol. 163^{v}.

475. **Merton College 193** (C.2.2) Saec. xv$\frac{2}{4}$
Peter of Herenthals, Comm. on the Psalter (250 × 165)

Fine borders, initials. Penwork initials.

Prov. Given by Robert Elyot, fellow of All Souls College, Oxford, vice-provost of Eton College (d. 1498).

Powicke, *Medieval Books*, 223, no. 1131; Emden, 638.

476. **Magdalen College Lat. 190** Saec. xv$\frac{2}{4}$
Martinus Polonus, Chronicle. Ranulf Higden, Polychronicon with continuation to 1377 (315 × 210)

Good borders (mutilated), initials. Good penwork initials. Marginal drawings, fol. 228^{rv}. *Mappa mundi*, fol. iii^{v}.

M. Destombes, *Mappemondes*, no. 47. 8; J. W. Taylor, *The Universal Chronicle of Ranulf Higden*, 1966, 116, 157, 178.

477. **New College 39** Saec. xv$\frac{2}{4}$
Gregory the Great, Moralia in Job. Bks. XVII–XXXV (355 × 250)

Fine border, initials. Good penwork initials.

Prov. Given by William Say, fellow, dean of St. Paul's, London (d. 1468), fol. 2.

Emden, 1650; N. R. Ker, 'The English MSS. of the Moralia of Gregory the Great', *Kunsthistorische Forschungen Otto Pächt zu ehren, eds.* A. Rosenauer, G. Weber, 1972, 79, 84–5, no. 47; Hunt, *Medieval library*, 327.

478. **All Souls College 14** Saec. xv¹
Hugh of St. Victor, Comm. on Ecclesiastes. Rhabanus Maurus, Comm. on Maccabees (269 × 163)

Good initials. Penwork initials, cadels.

Contemporary binding of white leather on boards.

Prov. Given by dominus William Folowys, grammar-master at Evesham (d. 1483), fol. 2.

Emden, 675.

479. **University College 71** Saec. xv¹
Septuplum (300 × 199)

Good initials. Penwork initials, cadels.

Contemporary binding of white leather on boards.

Prov. Given by William Hawkyns, fellow, 1527, fol. 14ᵛ.

Hunt, *Manuscript*, 16 n. 4.

480. **St. John's College 113** Saec. xv¹
English (?)
Anonymous, Comm. on Aristotle, Problemata (240 × 165)

Diagrams, penwork initials.

Prov. Clement Burdet, fellow of Queen's College (?), rector of Englefield, Berks., 1542–60/1, fol. 1. Given by John Stonor of North Stoke, Oxon., fol. 1ᵛ, saec. XVII in.

481. **All Souls College 83** Saec. xv¹
Seneca, Tragedies (260 × 170)

Initials.

Medieval boards with modern white leather.

482. **All Souls College 29,** fols. 195–234 Saec. xv¹
Rhabanus Maurus, in laude Sanctae Crucis (390 × 283)

Diagrammatic tables (incomplete).

Prov. Thomas Winterborne, fellow, dean of St. Paul's, London (d. 1478), fol. 234ᵛ.

Emden, 2061.

483. **Merton College 265** (C.2.16) Saec. xv¹
Roger of Waltham, Compendium morale (210 × 145)

Good penwork border, initial.

Prov. Given by Thomas Bloxham, fellow, canon of Beverley (d. 1473), fol. i.

Emden, 205.

484. **Magdalen College Lat. 186** Saec. xv¹
Aegidius de Columna, Comm. on the Sentences. Bk. I (320 × 228)

Initial. Penwork initials.

485. **Balliol College 316B** Saec. xv¹
Boethius, de consolatione philosophiae. English translation by John Walton (258 × 162)

Borders, initials. Scribbles and rough sketches, fols. 95, 115ᵛ, saec. XVI (?).

M. Science, 'Boethius, *De consolatione philosophiae* translated by John Walton', *E.E.T.S.*, CLXX (1927), xvii.

486. **Corpus Christi College 124** Saec. xv¹
Raymond Lull, Liber de quinta essentia, etc. (192 × 138)

Good initial. Penwork initials.

Contemporary binding of white leather.

James, *Dr. John Dee*, 21.

487. **Corpus Christi College 241** Saec. xv¹
Anonymous, Casus summarii. Gregory IX, Decretals, etc. (167 × 115)

Drawings (partly coloured) added to a MS. of saec. XIV ex., fols. 1ᵛ–5. [Illus.]

Binding of faded pink leather on boards, saec. XV ex.

488. **Magdalen College Lat. 82** Saec. xv¹
Alan of Lille (?), Comm. on Ps.-Cicero, Rhetorica ad Herennium (277 × 172)

Initial.

G. Raymond de Lage, *Alain de Lille*, 1951, 17 n. 22.

489. **Lincoln College Lat. 88** Saec. xv¹
Sermons. On paper and parchment (365 × 252)

Good initial.

Brother Bonaventure, 'The teaching of Latin in late medieval England', *Medieval Studies*, XXIII (1961), 3, 14 n. 46, 20.

490. **Balliol College 287,** fols. 58–106ᵛ 1450
Guarino da Verona, Lexicon (295 × 205)

Good borders (mutilated). Written for Richard Bole in humanistic script by Theodericus Werken, 1450, fol. 106ᵛ. Cf. no. 447. For other MSS. see Index.

Prov. Richard Bole, fellow, archdeacon of Ely (d. 1477), fol. 1, 1ᵛ.

Weiss, *Humanism*, 96 n. 8; de la Mare, *A fragment*, 285 n. 3.

491. **Balliol College 127** *c.* 1450
Petrarch, Cicero, Poggio (265 × 120)

Good borders, initials. Humanistic script by Theodericus Werken, 1450, fol. 164. Cf. no. 447. For other MSS. see Index.

Prov. Written for Richard Bole, archdeacon of Ely (d.

1477), who gave it to the College, fol. 4^{v}. His motto 'Abide and Have' incorporated in the border, fol. 5.

Exh., *Italian Illuminated MSS.*, 1948, no. 93; Mynors, *Werken*, 101, 104; Exh., *Duke Humfrey*, 25; N. Mann, 'Petrarch's Role in Humanism', *Apollo*, XCIV, no. 115 (1971), 180, figs. 5, 6; *id.*, 'Petrarch manuscripts in the British Isles', *Italia medioevale e umanistica*, XVIII (1975), no. 154; de la Mare, *A fragment*, 285 n. 3, 9.

492. **New College 266** *c.* 1450
John Gower, Confessio amantis (third version) (330 × 220)

Fine miniatures (mutilated and defaced), borders, initials. Possibly by William Abell. Cf. nos. 495, 564, 578. [Illus.]

Binding of blind-stamped leather on boards.

Prov. John Cutt of Shenley, fol. 194, saec. XV ex. Thomas Martin, fellow (d. 1584), fol. 1.

G. C. Macaulay, 'The English works of John Gower', *E.E.T.S.* ex. ser. LXXXI (1900), II, 81–2; J. H. Fisher, *John Gower*, 1965, 125, 127, 352; J. J. G. Alexander, 'William Abell "lymnour" and 15th-century English Illumination', *Kunsthistorische Forschungen Otto Pächt zu ehren, eds.* A. Rosenauer, G. Weber, 1972, 168, no. 20.

493. **Merton College 247** (B.1.5) *c.* 1450
Pierre Bersuire, Reductorium morale (293 × 210)

Border, initial.

Prov. Given in 1493 by William Romsey, fellow (d. 1501 (?)).

Powicke, *Medieval Books*, no. 1104; Emden, 1606.

494. **St. John's College 293** *c.* 1450
Psalter (mutilated) (265 × 170)

Good miniatures, borders, historiated and other initials. Good Lombards. [Illus.]

Prov. Calendar points to London (translation of Edward the Confessor, 13 Oct.). William Smallwood, fellow of Magdalen College, 1553, and St. John's College 1557–66, fols. i, 8, etc. Purchased 1952.

495. **St. John's College 18** *c.* 1440–*c.* 1465
Rolandus Alixbonensis, Reductorium phisionomie (355 × 255)

Good miniature, border. Fine and other initials. By William Abell who was paid for illuminating the Eton College consolidation charter of 1446. Cf. nos. 492, 564, 578. [Illus.]

Prov. Henry Savile (d. 1617), fol. 1. Given by sir William Paddy, commoner, physician to James I (d. 1634), fol. 2.

A. G. Watson, *The Manuscripts of Henry Savile of Banke*, 1969, no. 3; J. J. G. Alexander, 'William Abell "lymnour" and 15th-century English Illumination', *Kunsthistorische Forschungen Otto Pächt zu ehren, eds.* A. Rosenauer, G. Weber, 1972, 167, no. 15.

496. **Balliol College 67B** Saec. XV med.
Franciscus de Mayronis, Comm. on the Sentences (incomplete) (400 × 275)

Fine borders, initials (mutilated), in the same style as MSS. Balliol 68, 70. Written for William Gray while in Italy 1445–53 by the scribe of MSS. Balliol. 68, 70 fols. 2–116^{v}. Cf. nos. 497, 498.

Prov. William Gray, bp. of Ely 1454–78, fol. 1^{v}.

Mynors, *Catalogue*, XXXV.

497. **Balliol College 68** Saec. XV med.
Franciscus de Mayronis, Comm. on the Sentences (incomplete) (405 × 296)

Good borders, initials. Penwork initials. Written for William Gray while in Italy 1445–53 by the scribe of MSS. Balliol 67B, 70. Cf. nos. 496, 498.

Prov. William Gray, bp. of Ely 1454–78, fol. 1^{v}.

Mynors, *Catalogue*, XXXV.

498. **Balliol College 70** Saec. XV med.
Franciscus de Mayronis, Flores Augustini (393 × 296)

Good borders, initials. Penwork initials. In the same style as MSS. Balliol. 67B, 68. Written for William Gray while in Italy 1445–53 by two hands changing on fols. 116^{v}–17. The first scribe also wrote MSS. Balliol. 67B, 68. Cf. nos. 496, 497.

Prov. William Gray, bp. of Ely 1454–78, fol. 1^{v}.

Mynors, *Catalogue*, XXXV.

499. **Balliol College 98** Saec. XV med.
Albertus Magnus, de sophisticis elenchis, etc. (mutilated) (400 × 270)

Good borders, initials. Penwork initials. Written by the Italian scribe of MS. Balliol 100 (no. 500), presumably during Gray's stay in Italy, 1445–53.

Prov. William Gray, bp. of Ely 1454–78, offset of his arms on fol. 2^{v}.

Mynors, *Catalogue*, XXXV.

500. **Balliol College 100** Saec. XV med.
Albertus Magnus, Comm. on Aristotle, Physics, etc. (incomplete) (394 × 278)

Good initials (mutilated). Penwork initials. Written by the Italian scribe of MS. Balliol 98 (no. 499), probably 1445–53.

Prov. William Gray, bp. of Ely 1454–78, offset of his arms on fol. 1ᵛ.

Mynors, *Catalogue*, xxxv.

501. **Lincoln College Lat. 81** Saec. xv med.
Johannes Parisiensis, Boraston, Peckham, etc. Misc. theological tracts (295 × 198)

Good initials. Penwork initials.

502. **Magdalen College Lat. 95** Saec. xv med.
Philip the Chancellor (?), Sermons (287 × 210)

Good borders (mutilated). Initials.

J. B. Schneyer, 'Sermones de tempore', *Repertorium d. Lateinisch. Sermones des Mittelalters* (Beiträge zum Geschichte d. Philosophie und Theologie des Mittelalters, Band XLIII, Heft 4), 1972, 825.

503. **Balliol College 216** Saec. xv med. Oxford (?)
Tabula septem custodiarum. Part I (Comm. on the Old Testament) (330 × 250)

Good borders, initials. Written for William Gray by Johannes Reynbold from Zierenberg in Hesse, a German scribe working in Oxford, colophon fol. 140ᵛ. Cf. no. 506. For other MSS. see Index. Companion volume to MS. Balliol 217 (no. 386) obtained from some other source.

Prov. William Gray, bp. of Ely 1454–78, fol. 1ᵛ.

Mynors, *Catalogue*, xlii.

504. **Balliol College 16** Saec. xv med. Oxford
Hugh of St. Cher, Comm. on Ezechiel (incomplete) (342 × 293)

Good border, initial. Penwork initials. By the same illuminator as MSS. Balliol 17 and 18 (nos. 505, 507). Written for William Gray by a probably German scribe working in Oxford who also wrote MS. Balliol 17.

Prov. William Gray, bp. of Ely 1454–78, fol. 1ᵛ.

Mynors, *Catalogue*, xxvi.

505. **Balliol College 17** Saec. xv med. Oxford
Hugh of St. Cher, Comm. on Jeremiah (342 × 298)

Good border, initials. Penwork initials. By the illuminator of MSS. Balliol College 16 and 18 (nos. 504, 507). Written for William Gray by a probably German scribe working in Oxford who also wrote MS. Balliol 16.

Prov. William Gray, bp. of Ely 1454–78, fol. 1ᵛ, his arms in the initial fol. 2.

Mynors, *Catalogue*, xxvi.

506. **Balliol College 261** Saec. xv med. Oxford
Roger of Waltham, Compendium morale (262 × 180)

Good border, good and other initials. Written by Johannes Reynbold from Zierenberg in Hesse, a German scribe working in Oxford, cf. no. 503. For other MSS. see Index.

Prov. Robert Thwaites, master, dean of Auckland, co. Durham (d. 1458), fol. 1ᵛ.

Mynors, *Catalogue*, xx–xxi.

507. **Balliol College 18** Saec. xv med. Oxford
Hugh of St. Cher, Comm. on Genesis (342 × 293)

Good borders, initials. Penwork initials. By the illuminator of MSS. Balliol College 16 and 17 (nos. 504, 505). [Illus.]

Prov. William Gray, bp. of Ely 1454–78, fol. 1ᵛ, his arms in the initial fol. 2.

Mynors, *Catalogue*, xxvi.

508. **St. John's College 194,** fol. 2 Saec. xv med. (?) Christ Church, Canterbury (?)
Gospels (164 × 127)

Drawing added to a Breton (?) MS. of saec. IX/X.

Prov. Cf. no. 1.

509. **University College 78B** Saec. xv med. York Minster
Missal. Use of York (334 × 220)

Good and other initials. Good penwork initials. Sketch of a figure, fol. 184ᵛ, saec. xv. [Illus.]

Contemporary binding of white leather on boards.

Prov. Calendar (feast of relics 19 Oct.) suggests York Minster. Obit of Margaret Haverham, 1532. Given to Cuckney church by William Sheppard, fols. 1, 184ᵛ. Given to the College by Thomas Walker, master (d. 1665).

W. Henderson, 'Missale ad usum Ecclesiae Eboracensis', *Surtees Society*, 59–60 (1874), vi, x; Hunt, *Manuscript*, 21.

510. **St. John's College 58** Saec. xv med. Oxford
De morte et conquesta, and other poems (303 × 205)

Diagrams, tables, tree of the virtues and vices.

Prov. Given by Gabriel Collinge, chaplain of the College, 1618, fol. 1.

511. **Balliol College 30** Saec. xv med. Oxford (?)
Thomas Docking, Comm. on St. Paul's Epistles (incomplete) (415 × 292)

Good borders, initials (mutilated). Written for William Gray by a Dutch scribe, Tielman Reynerszoon (ejaculation 'Maria hilff', fol. 308^{v}), probably in Oxford *c.* 1442. Cf. no. 443, 928.

Prov. William Gray, bp. of Ely 1454–78. Note by Thomas Gascoigne, chancellor of Oxford University (d. 1458), fol. 275.

M. B. Parkes, *English cursive bookhands 1250–1500*, 1969, pl. 24 (ii); Ker, *College Libraries*, 309 n. 81.

512. **Balliol College 207** Saec. XV med. Oxford (?)
William of Auvergne, de universo, etc. (incomplete) (465 × 311)

Fine borders, initials. Penwork initials. By the same illuminator as MS. Balliol 174 (no. 526). Written by the same scribe, Henry Mere, cf. no. 526.

Prov. William Gray, bp. of Ely 1454–78, offset of his arms on fol. 1^{v}.

Ker, *College Libraries*, 309 n. 80.

513. **Keble College 14** Saec. XV med.
Book of Hours. Use of Sarum (208 × 136)

Fine miniatures, borders. Fine and other initials. By two hands. Line fillers. Penwork initials. [Illus.]

Prov. Translation of St. Erkenwald suggests London. Sir John Richardson, explorer (d. 1865), fol. i. Bequeathed by Dr. Henry P. Liddon, canon of St. Paul's, London (d. 1890).

514. **Balliol College 236** Saec. XV med.
Ranulf Higden, Polychronicon to 1346 combined with John of Tynemouth, Historia aurea (460 × 300)

Good border, initials. Penwork initials.

Prov. Humphrey Stafford, first duke of Buckingham (d. 1460), uncle of William Gray, bp. of Ely 1454–78, who gave it to the College, fol. i^{v}.

V. H. Galbraith, 'Extracts from the *Historia aurea* and a French 'Brut' (1317–47)', *English Historical Review*, XLIII (1928), 205 n. 2.

515. **Balliol College 224A** Saec. XV med.
Richard Rolle, Various works (308 × 210)

Good borders, initials (mutilated). Border on fol. 5 added to a slightly earlier MS.

Prov. William Gray, bp. of Ely 1454–78, fol. 3^{v}.

516. **Magdalen College Lat. 23** Saec. XV med.
Terence, Comedies (242 × 178)

Fine historiated and other initials by two hands. Humanistic script in part by Thomas Candour, cf. no. 575. For other MSS. see Index. [Illus.]

Medieval binding of white leather on boards.

Prov. Thomas Ashwell, Carmelite friar (d. 1468), erased inscription fol. 155^{v}. Samuel Wrott, armiger, tenant of the College, fol. 2, saec. XVI.

Emden, App. 2158; Exh., *Duke Humfrey*, 32; N. Mann, 'Petrarch manuscripts in the British Isles', *Italia medioevale e umanistica*, XVIII (1975), no. 240; A. C. de la Mare, 'Humanistic hands in England', Exh., *Manuscripts at Oxford*, XXII, 96.

517. **Balliol College 300A** Saec. XV med.
John of Salisbury, Polycraticus (375 × 253)

Initials (mutilated). Written in humanistic script by Richard Bole, colophon fol. 179. For other MSS. see Index.

Prov. Richard Bole, fellow, archdeacon of Ely (d. 1477).

Mynors, *Catalogue*, xlix; Exh., *Duke Humfrey*, 25, 30.

518. **Corpus Christi College 218** Saec. XV med.
Robert Silk, Fasciculus morum, etc. (213 × 165)

Good and other initials, border. [Illus.]

519. **Corpus Christi College 56** Saec. XV med.
Ludolphus de Saxonia, de vita Christi (495 × 305)

Fine borders (mutilated). Fine initials by two hands.

Medieval binding of white leather on boards.

520. **Exeter College 27** Saec. XV med.
John Lathbury, Comm. on Jeremiah, Lamentations (348 × 240)

Fine borders, initials. Penwork initials.

521. **Lincoln College Lat. 28** Saec. XV med.
Bernard of Clairvaux, de compassione B.V.M. (200 × 124)

Initials.

522. **Lincoln College Lat. 56** Saec. XV med.
Robert Grosseteste, Dicta, Sermons, etc. (265 × 165)

Fine borders, initials. Penwork initials.

Medieval binding of white leather on boards.

Prov. 'M. W. Rede', fol. 159. Bequeathed by William Rede, vicar of St. Mary Magdalen's, Oxford (d. 1469), to John Asheby for his lifetime and after his decease to the College.

H. E. Salter, 'Registrum Cancellarii Oxoniensis. II', *Oxford Historical Society*, XCIV (1932), 307; Emden, 56, 1560.

523. **All Souls College 345,** fols. 1, 76 Saec. xv med.
Breviary (fragment) (297 × 185)

Initials (rubbed).

524. **Keble College Roll 1** Saec. xv med.
Pontifical Benedictions (roll) (3140 × 210)

Good border, historiated initial (rubbed), other initials. Penwork initials.

Prov. From the archives of the Towneley family. Bequeathed by Canon C. E. Brooke (d. 1911).

525. **Magdalen College Lat. 147** Saec. xv med.
Ranulf Higden, Polychronicon with continuation to 1381, and other chronicles (320 × 250)

Fine border, initials, good penwork initials.

Prov. St. Sepulchre, Newgate.

J. W. Taylor, 'The development of the Polychronicon continuation', *English Historical Review*, LXXVI (1961), 24, 32, 35–6; *id.*, *The Universal Chronicle of Ranulf Higden*, 1965, 113, 123 n. 2, 3, 157, 178, 180.

526. **Balliol College 174** Saec. xv med.
William of Auvergne, de fide et legibus, etc. (460 × 310)

Fine borders (mutilated), initials in Franco-Flemish style. Penwork initials. Written by Henry Mere, 'quod h' and a paraph identifiable as his signature, fol. 155v. Cf. no. 512. [Illus.]

Prov. William Gray, bp. of Ely 1454–78, offset of his arms, fol. 1v.

Ker, *College Libraries*, 309 n. 80.

527. **New College 114** Saec. xv med.
Duns Scotus, Comm. on the Sentences (415 × 295)

Fine border, initial. Good penwork initials. Marginal drawings (catchwords) are probably an addition of saec. xv ex. (?). Written by a foreign scribe, perhaps German or Flemish, possibly working in England. [Illus.]

528. **Balliol College 316A** Saec. xv med.
Boethius, de consolatione philosophiae. English translation by John Walton (250 × 107)

Good borders, initials.

Prov. Anne Gifford, wife of Richard Gifford of Mottisfont, armiger, 1656, fol. 1v.

M. Science, 'Boethius, *De consolatione philosophiae* translated by John Walton', *E.E.T.S.* CLXX (1927), xii.

529. **Balliol College 349** Saec. xv med.
Willelmus de Lanicia, Dieta salutis, etc. (273 × 202)

Good initial. Penwork initials. Rough sketch, first pastedown.

Binding of faded red leather over bevelled boards, saec. xv.

Prov. Bequeathed by Dr. George Coningesby, antiquary, 1766, fol. 49.

530. **Corpus Christi College 88** Saec. xv med.
Petrarch, Secretum. Poggio, De vera nobilitate (220 × 150)

Good initial.

Prov. Motto '. . . audi vide tace', fol. 1. Given by the founder, Richard Foxe, bp. of Winchester 1501–28.

Emden, 717; N. Mann, 'Petrarch manuscripts in the British Isles', *Italia medioevale e umanistica*, XVIII (1975), no. 238.

531. **Corpus Christi College 236** Saec. xv med.
Richard Rolle, Incendium amoris, etc. English translation by Richard Misyn, *c.* 1435 (310 × 200)

Fine border, initials. [Illus.]

Contemporary binding of white leather on boards.

Prov. Dr. John Dee (d. 1608).

R. Harvey, 'The Fire of Love. The Mending of Life', *E.E.T.S.*, or. ser., CVI (1896); James, *Dr. John Dee*, 24; Emden, 1286.

532. **New College 113** Saec. xv med.
Duns Scotus, Comm. on the Sentences (409 × 290)

Good and other initials.

Prov. Given by William Warham, fellow, archbp. of Canterbury 1503–32.

Hunt, *Medieval library*, 330.

533. **New College 8–13** Saec. xv med.
Nicholas de Lyra, Comm. on the Bible (625 × 325)

Good borders, fine and other initials (mutilated). Line-fillers. Diagrams (vol. II).

534. **Merton College 179** (A.1.7) Saec. xv med.
Hugh of St. Cher, Comm. on the Apocalypse (280 × 200)

Fine border, initial. Penwork initial. Written by W. Breton, 1449, fol. 158v. [Illus.]

Contemporary binding of white leather on boards.

Prov. Given in 1493 by Robert Aubrey and Robert Field, fellow of Lincoln College, 'ex procuratione' of John Goodhow, fellow of Merton College in 1490, fol. i.

Emden, 682.

535. **Christ Church 151** Saec. XV med.
Boethius, de consolatione philosophiae. English translation by John Walton (217 × 160)

Border, initials.

Remains of original binding.

536. **Magdalen College Lat. 157** Saec. XV med.
Thomas Netter of Walden, de septem sacramentis (incomplete) (400 × 250)

Fine borders, initials. Penwork initials. [Illus.]

537. **Merton College 49** (C.1.10) Saec. XV med.
Hugh of St. Victor, Augustine, Ambrose, etc. (210 × 140)

Fine border, initial. Penwork initials.

538. **Magdalen College Lat. 158–9** Saec. XV med.
Robert Holcot, Reductorium morale (390 × 270)

Fine borders, initials. Good penwork initials by two hands.

539. **Magdalen College Lat. 188** Saec. XV med.
Institutiones linguae gallicane. In Latin, English, and French (incomplete) (320 × 210)

Initial. Penwork initial.

Contemporary binding of white leather on boards.

Prov. Regula 78: 'De A quando hoc signum the non subsequitur ut a monsieur le comte doxenforde de monsieur John de Waynflete' (possibly brother of William, bp. of Winchester, d. 1479), fol. 4.

540. **New College 144** Saec. XV med.
Hugh of St. Victor, Misc. works (297 × 190)

Good borders, fine and other initials.

541. **Merton College 26** (B.3.4) Saec. XV med.
Jerome, Misc. works (mutilated) (390 × 253)

Fine borders, initials. Penwork initials.

Prov. Arms of FitzJames of Dorset and Somerset, fol. 6.

Powicke, *Medieval Books*, no. 1182.

542. **St. John's College 132** Saec. XV med.
John Arderne, Practica chirurgiae (225 × 153)

Drawings, diagrams.

Prov. Sir William Paddy, commoner, physician to James I, 1634, fol. 1.

543. **New College 281** Saec. XV med.
Ptolemy, Almagest (300 × 200)

Fine border, initials added for John Farley to a MS. of saec. XIII med. with tables and diagrams. His annotations in humanistic script, fols. 8^{v}, 9, etc.

Emden, 667; Hunt, *Medieval library*, 334.

544. **New College 310** Saec. XV med.
Book of Hours. Use of Sarum (205 × 150)

Fine border (rubbed). Fine and other initials.

Medieval binding of brown leather on boards.

545. **Oriel College 5** Saec. XV med.
Seneca, Letters (320 × 220)

Good border, initials.

Prov. 'S.W., 1601', fol. 1. Given by Abel Gower, fellow 1585–1616, fol. 1.

546. **University College 8** Saec. XV med. English (?)

Book of Hours. Use of Sarum (140 × 98)

Miniatures (unfinished) in Flemish style inserted on separate leaves. Good penwork initials. [Illus.]

Remains of contemporary white leather binding pasted on original boards.

Prov. Arms of the Percy family on an added leaf, cut down, fol. 4^{v}. Given to Robert Lover, 10 July, 1549, by Greete de Medhurst in Sussex, fol. 107.

Frere, I, 149.

547. **St. John's College 187** Saec. XV med.
Office of the Holy Ghost (176 × 115)

Good borders, good historiated (one) and other initials. Penwork initials. Linefillers.

548. **University College 82** Saec. XV med.
Cartulary of Beverley Minster (345 × 262)

Fine historiated initials (drawings). Calligraphic interlace initials. [Illus.]

Metal cross, a roundel with a distich under horn cover on more recent binding of brown leather on pasteboard.

Prov. St. John's, Beverley, Yorks.

Hunt, *Manuscript*, 26–7.

549. **University College 85** Saec. XV med.
Alain Chartier, Quadrilogue. English translation (350 × 246)

Fine miniatures, borders, initials. Written with calligraphic flourishes by the same scribe as Bodl. MSS. Laud Misc. 570, Ashmole 764 and San Marino, Huntington Library MS. 932, the latter signed 'Ricardus Franciscus' and dated 1447. [Illus.]

Prov. Crest, two arms embossed vested azure, holding between the hands proper a garb or and the motto

'Oublier ne doit', pp. 1, 70. Verses in French by Michael Otheu, physician of Emperor Maxmilian II addressed to Anne, duchess of Somerset, added pp. 2, 18, 180, 1585.

Hunt, *Manuscript*, 26 n. 3; Pächt, Alexander, I, nos. 695, 726; K. L. Scott, 'A mid-fifteenth century English illuminating shop and its customers', *J.W.C.I.* XXX (1968), 170 n. 3; M. S. Blayney, *ed.* 'Fifteenth-century English translations of Alain Chartier *Le traité de l'Espérance* and *Le Quadrilogue invectif*', *E.E.T.S.* CCLXX (1974), ix, frontispiece.

550. **Magdalen College Lat. 96** Saec. XV med.
Sermones dominicales, etc. (350 × 240)

Good initial.

551. **University College 154** Saec. XV med.
Brut Chronicle to 6 Henry V (1418). In English (245 × 177)

Border, initials. Penwork initials.

Prov. Christopher Purpillus, 1604, pp. i, 241. Possibly given by Mary Langbaine, widow of Gerard Langbaine (d. 1692).

Hunt, *Manuscript*, 21.

552. *Vacat*

553. **Jesus College 107** Saec. XV med.
Gospels of St. Matthew and St. John with gloss (mutilated) (305 × 198)

Marginal drawings added to a MS. of saec. XII¾.

Medieval binding of white leather on boards, repaired.

Prov. Sir John Prise (d. 1555).

N. R. Ker, 'Sir John Prise', *The Library*, X (1955), 13 (J. 10).

554. **Balliol College 37** Saec. XV med.
Odo of Châteauroux, Comm. on the Psalter (264 × 190)

Border, initial (mutilated). Drawings of window tracery, fols. 1ᵛ–2 (inserted flyleaves), saec. XV (?).

Prov. Robert Rooke, fellow, rector of Ashstead, Surrey (d. 1458), fol. 2ᵛ.

B. Smalley, *The Study of the Bible in the Middle Ages*, 1952, 266 n. 3, 267; Emden, 1589.

555. **All Souls College 88** Saec. XV med.
Quaestiones morales (215 × 145)

Border, initial.

Contemporary binding of white leather on boards.

Prov. Walter Hopton, fellow (d. 1459), fol. 1ᵛ.

556. **Magdalen College Lat. 138** Saec. XV med.
Durandus, Rationale divinorum officiorum (390 × 265)

Fine and other initials. Penwork initials. Semi-humanistic script. Fols. 203–5ᵛ written by Thomas Candour. Cf. no. 575. For other MSS. see Index. [Illus.]

Contemporary binding of white leather on boards.

Prov. William de Laguna (d. 1475), fol. 207.

Emden, 1084, App. 2158.

557. **Balliol College 26** Saec. XV med.
Robert Holcot, Comm. on Minor prophets, etc. (265 × 188)

Initials.

Prov. William Gray, bp. of Ely 1454–78, fol. iᵛ.

558. **Trinity College 1** Saec. XV med.
Boethius, de musica (215 × 135)

Good and other penwork initials. Diagrams.

Prov. Given by the founder, sir Thomas Pope (d. 1559).

559. **Magdalen College Lat. 144** Saec. XV med.
Johannes de Imola, Comm. on the Clementines. On paper (400 × 270)

Good initials. Penwork initials.

Contemporary binding of white leather on boards, straps, clasps.

560. **University College 181** Saec. XV med.
Guillaume Deguilleville, Pilgrimage of the soul. English prose translation made in 1400 (280 × 182)

Fine borders, fine and other initials. Miniatures pasted in and removed.

Contemporary binding of white leather on boards, metal bosses on each cover, chemise.

Prov. Given in 1490 to dominus Henry Percy, prior of Newnham, Bedfordshire, by his predecessor, Johannis Renhall, fol. 40ᵛ. Robert Hendrington, 1577, fol. 1.

Hunt, *Manuscript*, 19.

561. **Lady Margaret Hall, Borough 5** Saec. XV med.
Book of Hours (fragment) (128 × 86)

Border, one good and other initials. Penwork initials. Prayer in English added, fol. 37ᵛ, saec. XV ex.

Prov. Scribbles in Italian with English translation, saec. XVI, fols. 34ᵛ, 40ᵛ. Thomas Maydwell, fol. 39ᵛ, saec. XVI. Narcissus Luttrell, 1693, fol. 1. Bequeathed by Miss C. M. Borough, 1973.

562. **Balliol College 131** Saec. XV med.
Humanistic miscellanea (258 × 158)

Fine initials imitating Italian style. Diagrams. Penwork initials. Written for William Gray in Italy (Rome ?), 1445–53, in humanistic script by three scribes: fols. 2–13^{v}, by the hand of MSS. Balliol 286 and 290; fols. 14–127^{v} by a similar hand; fols. 128–77, by the scribe of MS. Balliol 128. Cf. nos. 571, 614, 939. [Illus.]

Prov. William Gray, bp. of Ely 1454–78, fol. 1^{v}.

Mynors, *Catalogue*, xxxv; Exh., *Duke Humfrey*, 28.

563. **Balliol College 88** Saec. xv med.
Anonymous, de virtutibus et vitiis (465 × 290)

Good borders, initials. Penwork borders, initials.

Prov. Arms (tarnished): argent between two bendlets three (indeterminate charges) gules, fol. 3^{v}. Notes by Richard Bole, archdeacon of Ely (d. 1477), who gave it to the College, fol. 1^{v}.

564. **Jesus College 124** Saec. xv med.
Prayers and hymns to the Virgin (Roll) (Approx. 1500 × 225)

Fine miniatures, border, initials. By William Abell, cf. nos. 492, 495, 578. [Illus.]

Prov. Made for Margaret of Anjou, wife of Henry VI, represented with her arms (m. 1445). Possibly Ralph Sheldon, Weston, Warwicks. (d. 1684).

I. G. Philip, 'Sheldon Manuscripts in Jesus College Library', *B.L.R.* I (1939), 13; J. J. G. Alexander, 'William Abell "lymnour" and 15th-century English Illumination', *Kunsthistorische Forschungen Otto Pächt zu ehren, eds.*, A. Rosenauer, G. Weber, 1972, 167, no. 4; J. J. G. Alexander, 'Painting and manuscript illumination for Royal patrons in the later Middle Ages'. *English Court Culture in the Later Middle Ages, eds.* V. Scattergood, J. W. Sherborne, 1983, 152 n. 48.

565. **Jesus College 110** Saec. xv med.
Robert Grosseteste, de decem mandatis, de lingua (300 × 214)

Good border, initials. Penwork initials.

Medieval binding of white skin on boards, repaired.

Prov. Nicholas Kempston, rector of Bingham's Melcombe, Dorset (d. 1477), fols. 1^{v}, 3. M. Medehoke, 1501, fol. 3^{v}.

Emden, 1034.

566. **Balliol College 97** Saec. xv med.
Johannes Buridanus, Quaestiones (343 × 278)

Good borders, initials.

Prov. William Gray, bp. of Ely 1454–78, fol. ii^{v}, whose arms are in the initial, fol. 1.

567. **Balliol College 117** Saec. xv med.
Johannes Canonicus, Quaestiones on Aristotle, Physics, etc. (340 × 218)

Fine borders, initials (mutilated). Other initials.

Prov. George Neville, bp. of Exeter 1458–65, his insignia apparently painted on the binding since gone.

Emden, 1349.

568. **Merton College 59–64** (G.2.4–5; G.3.1–4) 1451–5
Duns Scotus, Comm. on the Sentences (385–405 × 253–63)

Fine borders, initials (mutilated). Good penwork initials, cadels. Written for Richard Scarborough, fellow, by Johannes Reynbold from Zierenberg in Hesse, a German scribe working in Oxford, cf. no. 503. For other MSS. see Index. [Illus.]

MSS. 61, 62 between original boards.

Prov. Bought from Richard Scarborough by Thomas Bloxham, fellow, canon of Beverley, who bequeathed it to the College (d. 1475).

Powicke, *Medieval Books*, 33, no. 984.

569. **New College 228** 1452
Aristotle, Ethics and Politics. Ps.-Aristotle, Economics. Latin translation by Leonardo Bruni (305 × 210)

Fine initials imitating Italian white-vine decoration. By the same hand as Bodl. Laud Misc. MS. 579. Fine marginal drawing, fol. 2. Written in humanistic script, 'Anno domini 1452. Per me Jo. R.', fol. 89^{v}, possibly John Russel, grammar master at Oxford in 1446–7. He also wrote Cambridge, Trinity College MS. R. 14.5, fols. 10–46, 50–66^{v}. [Illus.]

Prov. William Warham, fellow, archbp. of Canterbury 1503–32, fol. 1.

Emden, 1608–9; Exh., *Duke Humfrey*, no. 37, pl. IVb; Pächt, Alexander, 3, no. 1090; J. Soudek, 'Fifteenth-century humanistic bestsellers: the manuscript diffusion of Leonardo Bruni's annotated version of the (Pseudo) Aristotelian *Economics*', *Philosophy and Humanism. Renaissance Essays in Honor of Paul Oskar Kristeller, ed.* E. P. Mahoney, 1976, 130–7, 142, no. 27.

570. **New College 291** 1452
Bonaventura, Breviloquium (145 × 98)

Fine border, fine and other initials. Good penwork initials, marginal drawings (catchwords). Colophon (partly erased), fol. 177: '1452 inceptum V die Septembris et finitum ultimo die eiusdem'. Drawings added saec. XVI, fols. 1^{v}, 182. [Illus.]

571. **Balliol College 286** *c.* 1454
Flavio Biondo, Italia illustrata (300 × 210)

Fine initials imitating Italian white vine decoration. Written for William Gray in humanistic script by the same Italian scribe as MSS. Balliol 131¹, 290. Cf. nos. 562, 614. [Illus.]

Prov. William Gray, bp. of Ely 1454–78, fol. 1ᵛ, offset of his arms, fol. 2.

Exh., *Duke Humfrey*, no. 46.

572. **Merton College 65** (G.3.5) 1456
Duns Scotus, Quodlibets, etc. Franciscus de Mayronis, Various works, etc. (365 × 226)

Fine borders, initials (mutilated). Good penwork initials. Written by Johannes Reynbold from Zierenberg in Hesse, a German scribe working in Oxford, colophon fols. 110, 154ᵛ. Cf. no. 503. For other MSS. see Index.

Prov. Bequeathed by Thomas Bloxham, fellow, canon of Beverley (d. 1473).

Powicke, *Medieval Books*, no. 985; Emden, 205.

573. **Balliol College 189** *c.* 1458
King's Lynn
John Capgrave, Comm. on the Acts (335 × 220)

Good borders, initials (mutilated). Written by the author as a presentation copy for William Gray, signed 'Feliciter per Capgrave', followed by a trefoil-mark, fol. 177ᵛ.

Prov. Prefatory letter of dedication to William Gray, bp. of Ely 1454–78, fol. 2. Given by him to the College, fol. 1ᵛ.

P. J. Lucas, 'John Capgrave, O.S.A. (1393–1464), scribe and "publisher"', *Transactions of the Cambridge Bibliographical Society*, v (1969), 1–35, (6).

574. **Merton College 268** (O.1.10) 1458–9
Salisbury
Johannitius, Isagoge, Quaestiones on Galen (285 × 192)

Fine borders, initials (mutilated). Good penwork initials. Written in Salisbury by Herman Zurke of Greifswald, fols. 189ᵛ, 299ᵛ. He also wrote Bodl. MSS. Bodley 362, 361 and Laud Misc. 558. [Illus.]

Stamped leather binding over boards, label marks on the back cover. Mottoes 'Mon bien mondain' and 'Jesu Mercy Lady help', Salisbury, *c.* 1460.

Prov. Written for Gilbert Kymer, chancellor of Oxford University 1447–53, dean of Salisbury 1449 (d. 1463), who gave or bequeathed it to the old University Library. Given to Merton College by Robert Barnes, M.D. (d. 1604).

G. D. Hobson, *English binding before 1500*, 1929, 16, pl. 34; N. R. Ker, 'The old University Library', *B.L.R.* v (1954–6), 178–80; Pächt, Alexander, 3, no. 1059.

575. **Magdalen College Lat. 196** 1459 (?)
Boethius, de consolatione philosophiae (275 × 200)

Fine border, initials, imitating French ornament (mutilated). Written by Thomas Candour, 1459 (?), end pastedown (partly erased inscription). For other MSS. see Index. [Illus.]

Contemporary blind-stamped leather binding on boards.

Prov. John Neele, chaplain to William Waynflete, bp. of Winchester (d. 1498), end pastedown.

Emden, 1341, App. 2158; Exh., *Duke Humfrey*, no. 54; G. Pollard, 'The names of some English fifteenth-century binders', *The Library*, XXV (1970), 201; A. C. de la Mare, 'Humanistic hands in England', Exh., *Manuscripts at Oxford*, XXII. 3.

576. **Balliol College 258** 1461
Sallust, Catilina (180 × 123)

Fine borders, initials. Humanistic script by Richard Bole. Cf. no. 517. [Illus.]

Prov. Richard Bole, fellow, archdeacon of Ely (d. 1477), fol. 1ᵛ.

Exh., *Duke Humfrey*, 25, 30.

577. **Balliol College 34** 1461
Thomas Ringestede, Comm. on Proverbs (375 × 253)

Good initials. Penwork initials, line-fillers. Written in London in humanistic script by Theodericus Werken and, in part, by Richard Bole, colophon fol. 254ᵛ. Cf. nos. 447, 517. For other MSS. see Index.

Prov. Probably given by Richard Bole, fellow, archdeacon of Ely (d. 1477).

Mynors, *Werken*, 101, 104; Exh., *Duke Humfrey*, no. 52, pl. VII; A. C. de la Mare, *A Fragment*, 286 n. 11.

578. **Trinity College 75** 1461
St. Mary's, Abingdon
Missal. Benedictine use. (pars hiemalis, incomplete) (358 × 242)

Fine borders, historiated and other initials (partly defaced). By William Abell who was paid for illuminating the Eton College consolidation charter of 1446. Cf. nos. 492, 495, 564. [Illus.]

Prov. Arms of Abingdon abbey (marginal sketch), fol. 6ᵛ. The 'pars aestivalis' of the MS. written in 1461 for Abbot Ashenden of Abingdon is now Bodl. MS. Digby 227. Given by the founder, sir Thomas Pope (d. 1559).

Millar, II, 40. 92, no. 308b; F. Wormald, 'English

Benedictine Calendars after A.D. 1100', *H.B.S.* LXXVII (1938), 15; Rickert, *Painting in Britain*, 185, 249 n. 9; Pächt, Alexander, 3, no. 1065; J. J. G. Alexander, 'William Abell "lymnour" and 15th-century English Illumination', *Kunsthistorische Forschungen Otto Pächt zu ehren, eds.* A. Rosenauer, G. Weber, 1972, no. 18; Morgan, *Gothic MSS.* 246; R. Marks, N. Morgan, *The Golden Age of English Manuscript Painting*, 1981, 119, pl. 40.

579. **Balliol College 204** *c.* 1461
Oxford (?)

Duns Scotus, Ordinatio IV (Opus oxoniense) (430 × 295)

Fine border, initial. Penwork initials. One of a set, cf. MSS. 202, 203. Written for Gray by Johannes Reynbold from Zierenberg in Hesse, a German scribe working in Oxford, colophon fol. 279. Cf. nos. 503, 580, 613. For other MSS. see Index. [Illus.]

Prov. William Gray, bp. of Ely 1454–78, fol. i^{v}, his arms in the initial, fol. 1.

Mynors, *Catalogue*, xlii; Ker, *College Libraries*, 308 n. 78, 82.

580. **Balliol College 206** 1462–3
Oxford (?)

Duns Scotus, Comm. on the Sentences (Opus parisiense) (438 × 299)

Good borders, initials (mutilated). Good penwork initials. By the same illuminator as nos 579, 613. Written for William Gray by the same German scribe working in Oxford, Johannes Reynbold from Zierenberg in Hesse, colophon, fols. 41^{v}, 269^{v}. Cf. no. 503. For other MSS. see Index.

Prov. William Gray, bp. of Ely 1454–78, fol. 1.

Mynors, *Catalogue*, xlii; Ker, *College Libraries*, 308 n. 78, 82.

581. **New College 288** *c.* 1463

Thomas Chaundler, Collocutiones and Allocutiones on William of Wykeham, etc. (295 × 210)

Fine borders, initials. Good miniatures (coloured drawings). Humanistic script by John Farley. [Illus.]

Prov. Presentation copy sent to Thomas Bekyngton, fellow, bp. of Bath and Wells (d. 1465) who gave it to Wells Cathedral.

T. Kirby, 'On some fifteenth-century drawings of Winchester Cathedral, New College, Oxford, etc.', *Archaeologia*, LIII, 2nd ser., (1892), part 1, 232–99, pls. XIV–XVII; M. R. James, *The Chaundler MSS.* (Roxburghe Club), 1916, passim; J. W. Bennett, 'Andrew Holes . . .', *Speculum*, XIX (1944), 314–35; R. Weiss, 'New light on Humanism in England during the fifteenth century', *J.W.C.I.* XIV (1951), 21; Emden, 399; Weiss, *Humanism*, 79 n. 9, 135 ns. 1–10 and passim; Exh., *Duke Humfrey*, no. 36; Hunt, *Medieval Library*, 325 n. 25.

582. **All Souls College 15** 1463

Johannes de Turrecremata, Comm. on the Psalter (235 × 160)

Fine border. Fine and other initials. Penwork initials. 'Ap Howell scripsit', fol. 214^{v}. [Illus.]

Medieval binding of white leather on boards.

Prov. Given by Robert Honywood, fellow (d. 1522), first pastedown.

Emden, 957.

583. **Balliol College 291** 1464–5

Duns Scotus, Quaestiones (432 × 275)

Fine borders, initials (mutilated). Written for William Gray by Johannes Reynbold from Zierenberg in Hesse, a German scribe working in Oxford, fols. 78, 229. Cf. no. 503. For other MSS. see Index.

Prov. William Gray, bp. of Ely 1454–78, fol. 1.

Mynors, *Catalogue*, xlii.

584. **Merton College 150–3** (M.3.1–4) Before 1466

Hugh of St. Cher, Comm. on the Old Testament (Pentateuch-Psalter) (417 × 280)

Borders. Fine and other initials. Penwork initials. Written by Colyngbourne, vol. IV, fol. 198^{v}.

Vol. IV in contemporary binding of faded pink leather on boards.

Prov. Given in 1466–8 by Henry Sever, warden 1455–71, vol. I, fol. i^{v}.

Powicke, *Medieval Books*, no. 947.

585. **Corpus Christi College 207** *c.* 1467–9
London or Westminster

Genealogical Chronicle of the Kings of England. In English (415 × 235)

Good miniature, diagrams, initials. Roll chronicle designed to be made up into codex form. By the same illuminator and scribe as MS. Brasenose College 17 (no. 624) and closely related to Bodl. MSS. e Mus. 42 and Lyell 33. Includes only two of Edward IV's children. [Illus.]

Blind-stamped leather binding by John Reynes, London, *c.* 1510–20.

A. C. de la Mare, *Catalogue of the Collection of Medieval Manuscripts bequeathed to the Bodleian Library, Oxford, by James P. R. Lyell*, 1971, 82, pl. VI (2).

586. **Exeter College 51–67** Before 1469 Oxford

Hugh of St. Cher, Comm. on the Bible (408 × 272)

Fine borders, fine and other initials (mutilated). By two hands. Cadels. Written for Roger Keys by William Salomon, colophon MS. 53, fol. 272, 1457, etc. He also wrote Exeter College MS. 68 (no. 593). [Illus.]

Prov. Arms of Roger Keys, fellow and warden of All Souls College, Oxford, archdeacon of Barnstaple (d. 1477), MS. 51, fol. 1, etc. Notice of his gift to Exeter College, 1469–70, MS. 67, fol. 1.

Ker, *College Libraries*, 309 nos. 79, 82.

587. **Jesus College 114** 1468–76

Genealogical Chronicle of the Kings of England to Edward IV, etc. (310 × 263)

Good miniatures (coloured drawings). Initial, marginal drawings, tables. Written by Thomas Haselden, B.A. 1476, fol. 25ᵛ. Roll bound up into codex form. [Illus.]

Prov. Thomas Wilkins, rector of Llanmare, co. Glamorgan (d. 1698), fol. 1.

Emden, 883; K. L. Scott, 'John Lydgate's Lives of Saints, Edmund and Fremund', *Viator*, XIII (1982), 347–9 n. 50.

588. **Merton College 89** (F.1) *c.* 1470–80

Raymond Lull, Six treatises (305 × 215)

Fine border, initial. Penwork initials. Written in humanistic script by John Pacy, who also wrote and signed Cambridge, Pembroke College MS. 235 in 1464.

Prov. Given by Richard FitzJames, Warden 1483–1507, bp. of London (d. 1522), his arms, fol. 1, probably an addition.

Emden, 692; Exh., *Duke Humfrey*, no. 89.

589. **Trinity College 46** After 1471 Sheen

Psalter (287 × 150)

Fine borders, fine and other initials in Franco-Flemish style. Good penwork initials. Written by Stephen Dodesham, monk of Sheen (d. 1481/2), fol. 167ᵛ, who was not at Sheen before 1471. [Illus.]

Prov. Sheen Charterhouse, Surrey. Given in 1630 by Ralph Kettell, president of the College 1599–1643, fol. 1.

590. **St. John's College 23** *c.* 1473–5

Genealogical Chronicle of the Kings of England ending in 1453, with continuation to Edward IV and later additions (340 × 220)

Good miniature, diagrams, initial. Same iconography and style as Bodl. MSS. e Mus. 42 and Lyell 33. Roll bound up into codex form. Includes only six of Edward IV's ten children.

Prov. John Atkinson, commoner 1568–72, fol. 1

C. L. Kingsford, *English historical literature of the fifteenth century*, 1913, 164 n. 3.

591. **Lincoln College Lat. 77** Before 1474

Pliny the Younger, Letters. Ps.-Pliny, de viris illustribus (190 × 135)

Good initial. Humanistic script mainly by Robert Flemmyng (fols. 1–mid 155ᵛ).

Medieval binding of white leather on boards, rebacked.

Prov. Robert Flemmyng, dean of Lincoln (d. 1483), fol. 1ᵛ.

Weiss, *Catalogues*, 348 (41) n. 63; *id.*, *Humanism*, 104 n. 12; Exh., *Duke Humfrey*, no. 62.

592. **Balliol College 81** 1475 (?)

Alexander Carpenter, Destructorium vitiorum (385 × 273)

Good historiated border (drawing, unfinished). Good and other penwork initials. Marginal sketches, fols. 5, 109, 116ᵛ. '. . . Scriptum 1475', fol. 391ᵛ.

Prov. Robert Abdy, fellow (d. 1483).

Emden, 2, 360.

593. **Exeter College 68** Saec. xv_4^3 Oxford

Hugh of St. Cher, Sermons (408 × 272)

Good border, initial. Penwork initials. Written by William Salomon, who also wrote Exeter College MSS. 51–67 (no. 586).

Prov. Roger Keys, fellow and warden of All Souls College, Oxford, archdeacon of Barnstaple (d. 1477), cf. Exeter College MSS. 51–67.

594. **Magdalen College Lat. 211** Saec. xv_4^3

Theological florilegium (340 × 240)

Fine borders, initials. Good penwork initials. Added to a MS. written by a German scribe.

Medieval binding of white leather on boards (damaged).

595. **St. John's College 6** Saec. xv_4^3

John Lydgate, The Siege of Troy. In English (390 × 270)

Good penwork borders (rubbed), initials.

Prov. Arms of Galfridus Leyntale and the motto (?) 'Jaffie bien' added in the initial of the prologue, saec. XVI.

H. Bergen, 'Lydgate's Troy Book. A.D. 1412–20', *E.E.T.S.* I, ex. ser. CXXVI (1935), 36–8, no. 11.

596. **Exeter College 129** Saec. xv_4^3

John Lydgate, The Siege of Troy. In English (390 × 260)

Border, initial (rubbed). Penwork and calligraphic interlace initials. Cadels.

H. Bergen, 'Lydgate's Troy Book. A.D. 1412–20', *E.E.T.S.* I, ex. ser. CXXVI (1935), 42, no. 14.

597. **St. John's College 196** Saec. xv$\frac{3}{4}$
Psalter (160 × 99)

Fine borders, historiated and other initials. Good penwork initials. Line-fillers. [Illus.]

Prov. Calendar points to London (translation of Edward the Confessor; feast and translation of St. Erkenwald). Richard Jardfield, 1598, fol. i^{v}.

598. **Merton College 173–4** (K.3.1–2) Saec. xv$\frac{3}{4}$
Florus, Comm. on St. Paul's Epistles (370 × 250)

Fine borders, initials. Written by Gosmer, vol. I, fol. 135^{v}. [Illus.]

Prov. Given by Richard FitzJames, warden 1485–1507, bp. of London (d. 1522), his arms added in the borders.

Powicke, *Medieval Books*, no. 1184; Emden, 692.

599. **Magdalen College Lat. 214** Saec. xv$\frac{3}{4}$
Ludolphus de Saxonia, de vita Christi (462 × 310)

Fine border, initials. [Illus.]

Contemporary binding of faded pink leather on boards with chemise cut down, contents label under horn on the back cover.

600. **Magdalen College Lat. 213** Saec. xv$\frac{3}{4}$
John Gower, Confessio amantis (third version). In English and Latin (475 × 325)

Fine borders, initials. Good penwork initials. Fine calligraphic initials (explicits), marginal drawings (catchwords). (Illus.)

Prov. Given by magister Marchadin Hunnis, 28 Febr., 1610, fol. 1^{v}.

G. C. Macaulay, 'The English works of John Gower', *E.E.T.S.*, ex. ser. LXXXI (1900), clxiii.

601. **Queen's College 168** Saec. xv$\frac{3}{4}$ (after 1453)
Genealogical Chronicle of the Kings of England (Roll) (approx. 8100 × 368)

Miniature. Good border, initials. Roll chronicle including Edward, Prince of Wales (b. 1453).

602. **All Souls College 40** Saec. xv$\frac{3}{4}$ (after 1453)
Genealogical Chronicle of the Kings of England to Henry VI (470 × 325)

Good miniature, border, initial. Roll bound up into codex form. Includes Edward, Prince of Wales (b. 1453). [Illus.]

Prov. Narcissus Luttrel, 1687, front pastedown.

K. L. Scott, 'Lydgate's Lives of Sts. Edmund and Fremund', *Viator*, XIII (1982), 360.

603. **Trinity College 76** Saec. xv$\frac{3}{4}$
Gospel lectionary. Use of Sarum (295 × 195)

Good border, initials. Penwork initials.

Prov. Given by the founder, sir Thomas Pope (d. 1559), fols. 1, 84.

604. **Queen's College 207** Saec. xv$\frac{3}{4}$
Book of Hours (incomplete). In Latin and English (213 × 145)

Penwork initials (mutilated). Good pencil drawings added, fols. 166–75, saec. XV ex. [Illus.]

Fragment of original brown leather binding, boards gone.

Prov. John Purday, fol. 175, saec. XV.

605. **New College 280** Saec. xv$\frac{3}{4}$
Eutropius, Pomponius Mela, etc. (255 × 180)

Fine initials. Humanistic script by Thomas Candour, cf. no. 575. For other MSS. see Index. [Illus.]

Emden, App. 2158; Exh., *Duke Humfrey*, 32.

606. **Magdalen College Lat. 201** Saec. xv$\frac{3}{4}$
Valerius Maximus, Memorabilia, etc. (280 × 200)

Fine borders, initials.

607. **Magdalen College Lat. 89** Saec. xv$\frac{3}{4}$
Rhabanus Maurus, Anselm, Bonaventura, etc. Richard Ullerstone, Petitiones (290 × 210)

Good borders, initials.

Emden, 1929.

608. **Keble College 32** Saec. xv$\frac{3}{4}$
Breviary (incomplete). Augustinian use (208 × 138)

Fine borders, initials.

Prov. Augustinian priory at Walsingham. Richard Vowell, prior of Walsingham (1514–38), fol. 252^{v}. 'Quod Garnet', possibly Richard Garnet, canon of Walsingham (1514–48). William Bragge sale, Sotheby, Wilkinson and Hodge's, 7 June 1876, lot 43. Sir Thomas Brooke, Huddersfield (d. 1908). Bequeathed by his brother, Canon C. E. Brooke (d. 1911).

H. J. Warner, 'The Walsingham "Wishing Wells"', *Transactions of the Norfolk and Norwich Archeological Society*, VIII (1877), 51–66.

609. **New College 271,** fols. 157–72 Saec. xv$\frac{3}{4}$
Maffeo Vegio, Aeneid, Book XIII; Astyanax (287 × 210)

Good initial imitating Italian humanistic vine-stem style (cf. no. 437), possibly by Thomas Candour who also wrote the text in humanistic script, fols. 157–mid 167. Vine-stem initial added *c.* 1480 by the illuminator of Oxford, MS. Bodley 915 (cf. Pächt, Alexander, 3, no. 1088), fol. 167. Humanistic script, fols. 167–72.

Prov. Cf. no. 915.

Emden, App. 2158; Exh., *Duke Humfrey*, 32, no. 59.

610. **All Souls College 11** Saec. xv$\frac{3}{4}$
Missal (incomplete). Use of Sarum (440 × 300)

Good borders, initials (damaged by damp).

Original front pastedown now used as fly-leaf.

Prov. Dedication of the church of Tregare, Monmouthshire, in 1512 added in the calendar, fol. 4, saec. XVI.

611. **St. John's College 87** Saec. xv$\frac{3}{4}$
Terence, Comedies (270 × 180)

Good borders, initials. Penwork initials.

Prov. Ben Jonson, dramatist (d. 1637), fol. 1.

612. **Corpus Christi College 61** Saec. xv$\frac{3}{4}$
John Lydgate, Life of St. Edmund. In English (250 × 180)

Good border (rubbed), initials. Cadels.

Prov. Thomas Gorstelow, M.A., 1621, fol. 1.

613. **Balliol College 209** Saec. xv$\frac{3}{4}$
Duns Scotus, Quodlibets (incomplete) (428 × 290)

Fine borders, initials. Good penwork initials. By the illuminator of Balliol MSS. 204, 206 (nos. 579–80). Written for William Gray by Johannes Reynold from Zierenberg in Hesse, a German scribe working in Oxford, colophon, fols. 98^{v}, 113^{v}, 141^{v}. Cf. no. 503. For other MSS. see Index.

Prov. William Gray, bp. of Ely 1454–78.

Mynors, *Catalogue*, xlii.

614. **Balliol College 290** Saec. xv$\frac{3}{4}$
English (?)

Giovanni Tortelli, de orthographia (440 × 290)

Fine initials (mutilated) imitating Italian white vine decoration. Written for William Gray in humanistic script by the same Italian scribe as MSS. Balliol 131^{1} and 286, cf. nos. 562, 571. [Illus.]

Prov. William Gray, bp. of Ely 1454–78, fol. 1^{v}.

Exh., *Duke Humfrey*, 25, no. 46, pl. VI.

615. **New College 307** *c.* 1475–80
Albertus de Eyb, Margarita poetica (375 × 250)

Good historiated initial (partly defaced). Good penwork initials. Copied from a book printed at Rome in 1475. [Illus.]

Prov. Given by Robert Mason, fellow, archdeacon of Northumberland (d. 1480), fol. 1.

Emden, 1241; Hunt, *Medieval library*, 329.

616. **University Archives B.2** 1477–8
Oxford

Senior Proctor's book, etc. (330 × 207)

Border, historiated initials (rubbed and defaced). Other good initials. Penwork initials.

Prov. Arms of the University and royal arms in the initials, fols. 16^{v}, 22.

S. Gibson, *Statuta antiqua Universitatis*, 1931, xvi–xvii.

617. **Trinity College 10** *c.* 1484
Chronicle of England. Misc. texts (208 × 145)

Good initial (rubbed). Penwork initials. 'Summa chronologica ab adventu Britonum usque ad an. 1484', fol. 214^{v}.

618. **Merton College 23** (F.1) 1497–1503
Jerome, Comm. on Minor Prophets (393 × 258)

Fine historiated border (mutilated), fine and other initials. By two hands. The first Flemish or, possibly, English imitating Franco-Flemish ornament, fol. 1. The other hand English. Cf. Merton 25 (no. 619). Good penwork initials. [Illus.]

Contemporary binding of white leather on boards with chemise pasted down, rebacked.

Prov. Written in 1497 for Richard FitzJames, warden 1483–1507, bp. of Rochester 1497–1503, bp. of London (d. 1522), fol. 237^{v}, who is shown kneeling in the border, fol. 1.

Powicke, *Medieval Books*, no. 1181 n. 1; Emden, 692.

619. **Merton College 25** (B.3.3) 1497–1503
Jerome, Comm. on Isaiah (390 × 265)

Fine historiated border. Fine historiated and other initials. By two hands The first Flemish or, possibly, English imitating Franco-Flemish ornament, fol. 1. The other hand English. Good penwork initials. Companion volume to Merton College 23 (no. 618). [Illus.]

Contemporary binding of white leather on boards with chemise cut down, donor's label under horn on the back cover.

Prov. Given by Richard FitzJames, warden 1483–1507, bp. of Rochester 1497–1503, bp. of London (d. 1522), shown kneeling in the border, his arms impaling arms of Rochester, fol. 1.

Powicke, *Medieval Books*, no. 1179; Emden, 692.

620. **New College 326** Saec. XV ex.
John Gower, Confessio amantis (first version, revised). In English (345 × 245)

Good penwork initials. Lombards including human heads. Marginal sketches, fols. 41, 121, 141.

Prov. Arms of Mompesson of Bathampton (argent a lion sable charged on the shoulder with a martlet), fol. 1, and of Mompesson impaled by Drewe (ermine a lion passant gules), fol. 8, suggest that the book was made for John Mompesson (d. 1500) who married Isabel, daughter of Thomas Drewe. Given by sir Giles Mompesson to John Mompesson, 1650. Given to the College by Thomas Mompesson, fellow, in 1705, fol. 207ᵛ.

G. C. Macaulay, 'The English works of John Gower', *E.E.T.S.*, ex. ser. LXXXI (1900), cxlii; John Fisher, *John Gower*, 1965, 117, 125, 352; Hunt, *Medieval library*, 339–40.

621. **St. John's College 257,** fols. 1–221 Saec. XV ex.
Statutes of England from 25 Henry VI to 11 Henry VII (1495/6). In English (270 × 170)

Good borders, historiated and other initials. Belongs to a group of Statutes of which Harvard Univ., Houghton Libr. MS. Richardson 40 is the closest in script and illumination. [Illus.]

Prov. Royal arms in the border, fol. 86. Sir John Franklin of Gonalston, Notts., fol. 1. His arms on the binding, saec. XVII¹.

622. **Queen's College 357** Saec. XV ex.
Paraphrase on Ave Maria. Pilgrimage to Jerusalem, etc. In Latin and English (150 × 95)

Fine miniatures, borders. Fine historiated and other initials. Penwork initials. [Illus.]

Prov. John Bateman, University College M.A., 1600.

Exh., *World of the Bible*, Palestine Exploration Fund, Victoria and Albert Museum, London, 1965, no. 9.

623. **Trinity College 73** Saec. XV ex.
Book of Hours (incomplete). Use of Sarum (?) (218 × 153)

Fine borders, historiated (unfinished) and other initials. Fine penwork initials. [Illus.]

Prov. Tobie Langford, 1632, fol. 166.

624. **Brasenose College 17** Saec. XV ex.
Genealogical Chronicle of the Kings of England to the end of Edward IV's reign (d. 1483) (220 × 150)

Good miniature, border, initials, diagrams. By the same illuminator and scribe as Corpus Christi College MS. 207 (no. 585). Roll bound up into codex form. [Illus.]

Prov. Given by Peter Shakerley de Holme, armiger, 1724, fol. 1.

625. **All Souls College 26** Saec. XV ex.
Treatise on the Assumption of the Virgin. English translation by Robert Sukar of 1485. On paper (278 × 197)

Initials (coloured drawings).

Prov. Narcissus Luttrell, 1682, fol. 1, purchased at the sale of Richard Smith's library, London, 15 May, 1682.

626. **Brasenose College 16** Saec. XV²
Psalter (136 × 80)

Good borders (mutilated), historiated and other initials.

Prov. Arms in the borders: argent a lion salient azure, over all a fess sable with three balls or, erased and repainted, fols. 87, 151, with argent the Five Wounds bleeding gules. Bridgettine litany and a calendar from another MS. with added feasts of St. Bridget, fols. 22–253, addition of saec. XV ex.

627. **New College 167** Saec. XV and XVI
Medical treatises (250 × 182)

Marginal pen and pencil drawings added to a MS. of saec. XIII.

Prov. Henry Beaumond, fellow of Exeter College (d. 1415). Annotated by Richard Burgess, fellow 1486–1503. Dr. John Dee (d. 1608).

James, *Dr. John Dee*, 32; Hunt, *Medieval library*, 333.

628. **New College 61** Saec. XV/XVI
Nicholas Gorham, Commentaries (430 × 285)

Good initials, border. Penwork initials. Humanistic script.

Binding of white leather on boards.

Prov. William Warham, fellow, archbp. of Canterbury, 1503–32.

Emden, 1990.

629. **New College 43** Saec. XVI in.
Alexander Nequam, Comm. on the Song of Songs (380 × 250)

Good initials (unfinished). Penwork initials. Humanistic script.

Contemporary binding of blind-stamped leather on boards by John Reynes, London, 1510–20.

Prov. Arms of William Warham, fellow, archbp. of Canterbury 1503–32, fol. 3.

Emden, 1990; Hunt, *Medieval library*, 330; K. L. Scott, 'Lydgates Lives of Sts Edmund and Fremund', *Viator*, XIII (1982), 337 n. 8.

630. **New College 42** Saec. XVI in.
Bede, Comm. on the Song of Songs (325 × 235)

Good and other initials. Penwork initials. Humanistic script.

Contemporary stamped leather binding on boards.

Prov. William Warham, fellow, archbp. of Canterbury 1503–32, fol. 133.

Emden, 1990; Hunt, *Medieval library*, 330.

631. **New College 57** Saec. XVI in. Canterbury
Bede, Comm. on the Acts and the Apocalypse (282 × 190)

Good border with arms (mutilated). Good initials. Humanistic script.

Contemporary binding of blind-stamped leather on boards.

Prov. Copied from an exemplar in the library of Christ Church, Canterbury, fols. 2^v, 61^v. Arms of Christ Church, fol. 5.

Hunt, *Medieval library*, 330.

632. **St. John's College 173** Saec. XVI$^1_4$
Religious miscellanea. In Latin and English (190 × 125)

Initials. Penwork initials. Written by John Colman, fol. 44^v, with his monogram.

Prov. Ex libris of St. Mark's Hospital, Bristol, fol. 1, presumably contemporary. Given by Martin Okins, fellow 1610, fol. 2.

633. **New College 313** Before 1536
Robert Shirburn, bp. of Chichester 1508–36, Ordinationes episcopales (375 × 260)

Good borders, initials, imitating Flemish style (mutilated). Fine calligraphic interlace initials. [Illus.]

Heraldic blind-stamped calf binding with royal arms, London, saec. XVI (cf. J. B. Oldham, *English blind-stamped bindings*, 1952, pl. XLV, 735 a(1)). Metal boss with the title on the front cover, two metal bosses on the back cover.

Prov. 'Christopher Drone emebat pro Johanne Dele precii xxvj. s. viij sed vendidit'. Note on the admission to a prebend, 1571, fol. 1^v.

634. **New College 315** Saec. XVI^1
Registrum super scripturarum locos (355 × 220)

Fine calligraphic interlace initials. [Illus.]

635. **Lady Margaret Hall, Borough 21** 1559
Confirmation by letters patent of Elizabeth I of the grant made in 1551 by the Dean and Chapter of Christ Church Cathedral, Oxford, of the site of Clattercote Priory, Oxon., to Thomas Lee and Mary his wife (490 × 825)

Fine border, fine historiated initial (drawings). Cadels. [Illus.]

Prov. Bequeathed by Miss C. M. Borough, 1973.

636. **Christ Church 150** 1561
Book of Common Prayer (150 × 100)

Good border, initial.

Contemporary binding of white leather on boards.

637. **Corpus Christi College 153,** fols. 8–14 Saec. XVI ex.
John Dastyn, de virtute buffonis. On paper (224 × 125)

Good and other miniatures (drawings).

Prov. Brian Twyne, fellow (d. 1644), fol. 1.

638. **St. John's College 174** Saec. XVI^2
Caerlaverock poem. In French (190 × 140)

Good miniature (coloured drawing), coats of arms. Penwork initials. Authenticity testified by R. Glover, Somerset Herald (d. 1588), fols. iii^v, 33.

Prov. Philip Ayres, armiger, 1684, fol. 1.

A. R. Wagner, 'A catalogue of English mediaeval Rolls of Arms', *Aspilogia*, I (1950), 31, B.

IRISH MANUSCRIPTS

639. **Corpus Christi College 282** *c.* 1120–30
Missal (portable) (165 × 115)

Initials (damaged by damp). Irish script.

White leather binding on boards, original leather satchel.

Exh., *Latin Liturgical Manuscripts and Printed Books*, Bodleian Library, 1952, no. 38, pls. x, xi; F. Henry, G. L. Marsh-Micheli, 'A Century of Irish Illumination (1070–1170)', *Proceedings of the Royal Irish Academy*, LXII (1961–3), 102–3, 136–40, pls. XVII–XXI; F. Henry, *Irish Art in the Romanesque Period*, 1970, 48, 61–2, 120, 192, pls. 1, 4, 5, J.

640. **Corpus Christi College 129** Saec. XIV
Medical treatises (195 × 120)

Astrological tables, diagrams, initials (unfinished). Written in Irish minuscule.

FRENCH MANUSCRIPTS

Of the rather few early French manuscripts the unpublished Bible, All Souls MS. 4 (no. 647) is the most remarkable. It has no indication of its original provenance and so far no suggestions have been made and nothing similar has been found. It was in England already in the fifteenth-century. There are a number of thirteenth-century Bibles as is to be expected and of these Wadham College MS. 1 (no. 660) is important because dated 1244. Of other thirteenth-century manuscripts the New Testament in the French version of Guyart Desmoulins, Christ Church MS. 178 (no. 672) has not been reproduced before. Queen's College MS. 299 (no. 693) and Lincoln College Lat. 24 (no. 694) are fine quality Bibles from the same atelier as the Evangeliary of the Ste Chapelle in the British Library. The fragment of a Missal, Lincoln College Lat. 150 (no. 692) is an important recent accession. Balliol College MS. 2 is an interesting Bible, first because it has both French (no. 700) and Italian (no. 895) illumination in it, and secondly because there are preliminary marginal drawings for the French initials. Christ Church MS. 98 (no. 702) is an unpublished late thirteenth-century Psalter, probably from northern France.

Of the early fourteenth-century manuscripts Merton College MS. 123–6 (no. 716) is a four volume copy of Vincent of Beauvais' *Speculum historiale* of which volume III was written by John Wade, an English scribe working in Paris. St. John's College MS. 204 (no. 725), a Psalter, is also unpublished. Of the later fourteenth-century manuscripts St. John's College MS. 164 (no. 738) is a known manuscript containing astronomical treatises from the library of Charles V, and Keble College MS. 15 (no. 744) belongs with a group of Avignonese illuminated manuscripts. New College MSS. 185–6 (no. 755), Henri de Bohic, must have been a very impressive manuscript made no doubt in Paris in the early fifteenth century, but unfortunately its miniatures have been cut out. Keble College MS. 39 (no. 756) a Book of Hours illuminated by the Master of John, Duke of Bedford, has not yet entered the art historical literature in spite of its very high quality. The so-called Master of Sir John Fastolf who collaborated with the Bedford Master earlier in his career, is represented by two manuscripts, both known, the Missal, Keble College MS. 38 (no. 760) and the Prayers, St. John's College MS. 208 (no. 763).

The majority of the fifteenth- and sixteenth-century Books of Hours come from Keble College and include two fine examples by Jean Colombe, MSS. 42 and 41 (nos. 780, 781). Lincoln College Lat. 152 (no. 770) and Christ Church MS. 93 (no. 777) are both Hours from the workshop of Maître François, the latter with an interesting sequence of miniatures of St. Denis. Balliol College MS. 383 (no. 789) is a fine French Ovid of *c.* 1500 given to the College in 1839. An illuminator recently baptised the Master of Claude de France worked in two manuscripts, an Hours, Keble College MS. 43 (no. 795) and the splendid two volume Guyart Desmoulins, *Bible Historiale*, Corpus Christi College MSS. 385–6 (no. 796). The latter bought by its donor in Paris in 1720 is probably the most important unknown illuminated manuscript included in this volume and further research is needed into its provenance. There is a tradition that it belonged to François I. Corpus Christi College MS. 387 (no. 798), Book of Hours, is another important unknown manuscript a little later in date.

641. **St. John's College 150,** fols. 1–24 Saec. XI²
Guido of Arezzo, Micrologus, de arte musica, etc. (220 × 110)

Diagrams.

Prov. Augustinian priory of St. John, York, pressmark 'F', fol. 1ᵛ. John de Erghom, prior (d. 1390), fol. ivᵛ.

642. **Merton College 241** (C.1.16), fols. 2–17ᵛ
Saec. XI ex.
Normandy

Vita Malchi monachi (262 × 155)

One good initial. Bound up with other texts with arabesque initials, English, slightly later.

643. **Magdalen College Lat. 44** *c.* 1100
Horace, Various works (230 × 142)

Initials.

644. **Trinity College 20** *c.* 1100
Normandy (?)

Gospel of St. Matthew, Job, Apocalypse (incomplete), with gloss (255 × 162)

Good historiated and other initials. [Illus.]

B. Smalley, 'Les commentaires bibliques de l'époque romane', *Cahiers de Civilisation Mediévale*, IV (1961), 16–17, pl. IX.

645. **New College 252** Saec. XII[1]
French (?)

Cicero, Philippics, etc. (228 × 130)

Fine initial. [Illus.]

Prov. William Say, fellow, dean of St. Paul's, London (d. 1468), fol. 1.

A. C. Clark, 'The textual criticism of Cicero's Philippics', *Classical Review*, XIV (1900), 45–6, 250–1; Exh., *The Survival of Ancient Literature*, Bodleian Library, Oxford, 1975, no. 129.

646. **St. John's College 125** Saec. XII[1]
Ivo of Chartres, Panormia (227 × 130)

Initial.

Prov. Sir John White of Southwick, Hants., 1555, fol. 1. It may, therefore, have come from Southwick Priory like other of his gifts.

647. **All Souls College 4** Saec. XII$\frac{3}{4}$
Bible (mutilated and incomplete). 4 vols. (530 × 365)

Fine and other initials (Genesis initial pasted down and removed, fol. 3ᵛ). Arabesque initials, especially in vol. 2. Good initials, vol. 1, fols. 1–3, and penwork initials, vol. 2. fols. 118–124, added in England, saec. XV med. Marginal sketch, vol. 2, fol. 104ᵛ. [Illus.]

648. **Queen's College 386** Saec. XII$\frac{3}{4}$
French (?)

Augustine, Comm. on St. John's Gospel (420 × 285)

Good initials.

Original boards and sewing with later brown leather cover.

Prov. 'Hic est liber sancte Marie de M . . .' fol. 194ᵛ, saec. XIII (?). Possibly from a Cistercian house.

649. **Corpus Christi College 82** Saec. XII$\frac{3}{4}$
Quintus Curtius, Historia Alexandri magni. Julius Valerius, Epitome, etc. (460 × 345)

Fine arabesque and penwork initials.

Binding of blind-stamped leather on original boards, repaired, five metal bosses, clasps, by the 'Greyhound binder', *c.* 1500, Oxford.

Prov. List of contents, front pastedown, and annotations, fols. 113, 132, 190, by John Tiptoft, earl of Worcester (d. 1470). Given by the founder, Richard Foxe, bp. of Winchester 1501–28, fol. 1.

J. B. Oldham, *English blind-stamped bindings*, 1952, 21 n. 3, pl. XV; A. C. de la Mare, 'Humanistic hands in England', Exh., *Manuscripts at Oxford*, 99–100, XXII.7, figs. 70, 71; D. J. A. Ross, *Alexander Historiatus* (Warburg Institute Surveys, 1), 1963, 9, 86 n. 41, 90 n. 104, 101 n. 341.

650. **Merton College 200** (H.1.10) Saec. XII ex.
French (?)

Vocabularium biblicum (182 × 121)

Arabesque and penwork initials. Marginal sketch, fol. 43.

651. **Queen's College 318** Saec. XII[2]
Origen (incomplete) (285 × 217)

Fine and other initials.

Original boards.

652. **Balliol College 285,** fols. 68–187ᵛ Saec. XII/XIII
French (?)

Razes, Opera varia (305 × 210)

Fine initials. Penwork border, initials. Bound up with a cheiromantic tract in Anglo-Norman containing a left hand diagram, fol. 67ʳᵛ, saec. XIII[1] (cf. no. 677).

Fore-edge painting (traces of inscription).

Prov. William Reed, fellow of Merton College, bp. of Chichester (d. 1358), 'ex empto de bonis sibi datum per magistrum Nicolai de Sandvico', fol. 1ᵛ.

653. **Merton College 218** (N.3.1) *c.* 1200
French (?)

Galen, Opera varia (365 × 220)

Good initials. Good penwork initials.

Prov. Given by Richard FitzJames, warden (1483–1507), bp. of London (d. 1522), fol. 1ᵛ.

654. **St. John's College 34** *c.* 1200
French (?)

Peter Comestor, Historia scholastica (335 × 200)

Good initials. Penwork initials.

Prov. Given by sir John Lambe, commoner 1586 (?), registrar of Peterborough, fol. 2.

Exh., *Archbishop Laud*, Addenda.

655. **University College 112** *c.* 1200
Paris (?)

Peter Comestor, Historia scholastica (mutilated) (350 × 250)

Fine historiated and other initials. Good penwork initials. [Illus.]

Binding of white leather on boards, saec. XIV².

Prov. The will of Stephanus atte Roche, vicar of Fen Ditton, Cambs., 1381, formerly lining the upper cover of the binding, is now University College MS. 192, fol. 29.

Hunt, *Manuscript*, 15 n. 3.

656. **Pembroke College 8** Saec. XIII in.
Alexander of Tralles, Medical tracts (240 × 170)

Good initials (mutilated). Penwork initials.

Prov. Unidentified shelf-mark: 'S (?)·104·', saec. XV, fol. 1.

657. **Merton College 132** (L.1.4) Saec. XIII in. French (?)

Peter of Poitiers, Comm. on the Sentences. Simon of Tournai, Summa (345 × 210)

Fine initials. Penwork initials.

Medieval binding of boards, rebacked.

Prov. Dominican Convent, Oxford. Given by Thomas Bloxham, fellow, canon of Beverley (d. 1473), fol. 1ᵛ.

Emden, 205; Ker, *Medieval Libraries*, 142.

658. **Jesus College 38** Saec. XIII in. French (?)

Hugh of St. Victor, Comm. on Ps.-Dionysius, Areopagita (260 × 185)

Initial. Penwork initial.

659. **Pembroke College 13** Saec. XIII¼ French (?)

Johannes Platearius (?), de signis, causis et cura morborum (205 × 150)

Initial. Penwork initials. Diagrams.

660. **Wadham College 1** (A.5.2) 1244 Paris

Bible (230 × 155)

Fine historiated and other initials. Fine penwork initials. By the Gautier Lebaube atelier. Written in 1244 by 'Guillermus miles parisiensis . . .', fol. 434ᵛ. [Illus.]

Prov. Given in 1625 by William Boswell, fellow.

H. H. Glunz, *History of the Vulgate in England*, 1933, 273; Branner, 73 n. 8, 74, 213, figs. 148, 159; L. Eleen, *The illustration of the Pauline Epistles in French and English Bibles of the Twelfth and Thirteenth centuries*, 1982, 55 n. 51.

661. **St. John's College 9** Saec. XIII¹
Gospels with gloss (375 × 245)

Fine penwork initials.

Prov. Given in 1620 by William Laud, president, archbp. of Canterbury 1633–41, fol. 1.

Exh., *Archbishop Laud*, no. 9.

662. **New College 17** Saec. XIII¹
Bible (Leviticus-Deuteronomy) with gloss (350 × 227)

Fine historiated initials (mutilated) in two styles. Fine penwork initials. [Illus.]

Prov. Given by the founder, William of Wykeham, bp. of Winchester 1367–1404.

663. **Magdalen College Lat. 175** c. 1250 Paris

Isaac Judaeus, Misc. works. Other medical treatises (incomplete) (325 × 245)

Fine historiated and other initials with preliminary marginal sketches on fols. 142, 145, 177. Related in style to the Johannes Grusch atelier (cf. Branner, figs. 210–43). Fine penwork borders, initials. [Illus.]

664. **Trinity College 53** Saec. XIII med. Paris

Bible (317 × 206)

Fine historiated and other initials by the Bari atelier (cf. Bodl. MS. D'Orville 212, Branner, 229, fig. 300). [Illus.]

Prov. Identifiable as MS R.36 in the Syon abbey catalogue, saec. XV in. Given in 1569 by Francis Kettel of Kings Langley, Herefords., commoner of the College, fols. ivᵛ, 484.

M. Bateson, *Catalogue of the Library of Syon Monastery, Isleworth*, 1898, 175; Ker, *Medieval Libraries*, 187.

665. **New College 1** Saec. XIII med. Paris

Bible (305 × 215)

Fine historiated initials. [Illus.]

Prov. Given by the founder, William of Wykeham, bp. of Winchester 1367–1404. John Penyton, fellow 1495, fols. 409–10.

Emden, 1460.

666. **Balliol College 151** Saec. XIII med. Royaumont (?)

Bernard of Clairvaux, Letters (315 × 220)

Good penwork border, initials.

Prov. Contemporary *ex libris* (erased) of Royaumont abbey, Seine-et-Marne, fol. 199. William Gray, bp. of Ely 1454–78, fol. 1ᵛ.

667. **Balliol College 193** Saec. xiii med. Paris

Peter Lombard, Sentences (359 × 245)

Fine historiated initials (rubbed and mutilated). Penwork initials. [Illus.]

Prov. William Gray, bp. of Ely 1454–78, fol. 1^{v}.

668. **Corpus Christi College 17** Saec. xiii med.

Psalter (214 × 153)

Fine historiated initials. Fine line-fillers. Penwork initials. Cf. no. 371. [Illus.]

Prov. Calendar probably from the region of Autun (St. Leodegarius).

669. **Balliol College 8** Saec. xiii med. French (?)

Augustine, Comm. on Genesis. Isidore, Etymologiae (325 × 230)

Good diagrams, penwork initials.

Prov. Given by John Felton, rector of St. Mary Magdalen's, Oxford, 1420, fol. i^{v}.

Emden, 676.

670. **New College 174,** fols. 2–209^{v}, 236–301^{v} Saec. xiii med. French (?)

Justinian, Parvum volumen, with gloss of Accursius (415 × 235)

Fine miniatures, fine and other initials. Added to a MS. of saec. xii ex. Italian script and parchment. Cf. no. 269. [Illus.]

Prov. Lent by Thomas Bourton, fellow, canon of Salisbury 1416, to Thomas Clyffe, fellow, fol. 2^{v}.

Emden, 321, 448.

671. **Lincoln College Lat. 23** Saec. xiii$\frac{3}{4}$ Paris

Bible (265 × 165)

Fine historiated and other initials. Good penwork borders, initials. [Illus.]

Prov. Bequeathed by William Wheler to George Wheler, 1666, who gave it to the College, 1723, fol. 1.

672. **Christ Church 178** Saec. xiii$\frac{3}{4}$ Paris

Guyart Desmoulins, Bible historiale (New Testament). In French translation (342 × 258)

Fine borders, historiated initials by at least two hands. Penwork initials. [Illus.]

Prov. Jove Desplacet (?); S. Viriez (?), saec. xiv, fol. i; 1421 Valois, fol. i. Jhan Baptiste Deforests, saec. xv (?), fol. i. Possibly in Macon 1543 (note by Sebastien Tixier).

S. Berger, *La Bible française au Moyen Age*, 1884, 293, 413.

673. **Balliol College 253** Saec. xiii$\frac{3}{4}$ Paris

Aristotle, Isagoge, etc. (292 × 185)

Fine historiated and other initials. Penwork initials. In the style of the 'Corpus' atelier in Paris, cf. Branner. [Illus.]

Branner, 116, 117, 125 n. 27, cat. 235, figs. 347, 348.

674. **Jesus College 56** Saec. xiii$\frac{3}{4}$ Paris (?)

Bible (mutilated) (240 × 158)

Fine historiated and other initials. Penwork initials. Related in style to the Johannes Grusch atelier (cf. Branner, figs. 210–243). [Illus.]

675. **Exeter College 20** Saec. xiii$\frac{3}{4}$

Peter Lombard, Sentences, with gloss (mutilated) (400 × 257)

Fine initials. Penwork initials.

676. **Brasenose College 1** Saec. xiii$\frac{3}{4}$ Paris

Bible (mutilated) (390 × 250)

Fine historiated and other initials. Fine penwork borders. initials, line-fillers. Marginal drawings (catchwords). By the workshop of the Ste Chapelle Evangeliary (Bibl. Nat. lat. 17326). [Illus.]

Prov. Given by the founder, William Smyth, bp. of Lincoln (d. 1514).

Branner, 19, 125, 195, cat. 236, fig. 371; L. Eleen, *The illustration of the Pauline Epistles in French and English Bibles of the Twelfth and Thirteenth centuries*, 1982, 118–49.

677. **Balliol College 285,** fols. 189–220^{v} Saec. xiii$\frac{3}{4}$ Paris

Miscellaneous medical texts (305 × 210)

Good historiated initial. Bound up with earlier French (?) MSS., cf. no. 652.

Prov. Cf. no. 652.

678. **Keble College 20** Saec. xiii$\frac{3}{4}$ Paris

Bible (186 × 122)

Fine historiated and other initials. Penwork borders, initials. By the 'Aurifaber' atelier (cf. Branner, 109 ff., figs. 306–337). Possibly by the same hand as MS. Keble College 26, cf. no. 708. [Illus.]

Prov. Bequeathed by Dr. Henry P. Liddon, canon of St. Paul's, London (d. 1890).

M. B. Parkes, 'Punctuation or Pause and Effect', *Medieval Eloquence, ed.* J. J. Murphy, 1978, 134.

679. **Balliol College 277,** fols. 3–98^{v} Saec. XIII$\frac{3}{4}$
Aristotle, Metaphysics, etc. (incomplete) (340 × 220)

Fine initials. Fine penwork initials.

Prov. Pledged by magister Robert Roke in the Turville chest, Oxford University, between 1426–8, fols. 193, 193^{v}. Given by him to the College, fol. 2^{v}.

Aristoteles Latinus, I, 407, no. 357; Emden, 1589.

680. **Balliol College 250** Saec. XIII$\frac{3}{4}$
Aristotle, Rhetorical works (315 × 210)

Good borders, initials. Good penwork borders, initials. Marginal drawings by the scribe.

681. **Magdalen College Lat. 1** Saec. XIII$\frac{3}{4}$
Bible (142 × 100)

Good historiated and other initials.

Prov. Thomas Tresham, 1600, fol. 459^{v}. Given by sir Arthur Throckmorton, 1626, fol. 1.

682. **New College 26** Saec. XIII$\frac{3}{4}$
Gospels with gloss (440 × 300)

Fine borders, historiated and other initials.

Prov. Unidentifiable coats of arms on gilt foredges (rubbed).

683. **Jesus College 59** Saec. XIII$\frac{3}{4}$
Bible (155 × 106)

Good and other initials. Penwork initials.

Prov. V. Young, first pastedown, saec. XVI (?).

684. **Merton College 271** (O.1.3) Saec. XIII$\frac{3}{4}$
North France

Aristotle, de animalibus (304 × 205)

Fine borders, historiated and other initials (partly defaced). Preliminary marginal drawings. Good penwork borders, initials. [Illus.]

Powicke, *Medieval Books*, no. 392; Randall, *Images*, fig. 405.

685. **St. John's College 123** Saec. XIII$\frac{3}{4}$
Bible (225 × 150)

Fine historiated and other initials. Penwork initials.

Prov. Charles Margus, London, 1729, with his arms, first pastedown.

686. **Merton College 237** (H.2.4) Saec. XIII$\frac{3}{4}$
Sermons (245 × 170)

Good initial. Penwork initials.

Prov. Bought from Thomas Trillek, bp. of Rochester (d. 1372), by William Reed, bp. of Chichester (d. 1385), who bequeathed it to the College, fol. 1.

687. **Merton College 206–7** (N.2.3–4) Saec. XIII$\frac{3}{4}$
Bible (Genesis-Psalms) with gloss of Hugh of St. Cher (375–80 × 250)

Fine historiated and other initials. Fine penwork initials. Rough sketch of the Crucifixion prefacing the Gospels in vol. II. [Illus.]

Powicke, *Medieval Books*, no. 1197.

688. **Balliol College 195** Saec. XIII$\frac{3}{4}$
Peter Lombard, Sentences (405 × 266)

Fine historiated and other initials. Penwork initials. [Illus.]

Prov. Pledged in the New University chest, Oxford, by Richard Bosville, 1342, fol. 204^{v}; by magister Philipp de Wythemere (?), 1348, fol. 204, 204^{v}. Robert Pollan, fellow, vicar of Penistone, Yorks. (d. 1459), fols. 1^{v}, 203.

Emden, 1525, 2130.

689. **Trinity College 67** Saec. XIII$\frac{3}{4}$
French (?)

Aristotle, Misc. works (285 × 220)

Good border, initial (damaged by damp). Penwork borders, fine and other initials. Diagrams. Pen sketch, fol. 252^{v}, saec. XV (?).

Prov. Ex libris of Chertsey abbey, Surrey, fol. 251^{v}, seac. XIV (?). Given by the founder, sir Thomas Pope (d. 1559), fols. 2, 251^{v}.

690. **Merton College 222** (N.3.5) Saec. XIII$\frac{3}{4}$
Johannitius, Isagoge and other medical works (306 × 197)

Fine historiated and other initials. Penwork initials [Illus.]

Prov. Given by John Burbach, fellow, rector of St. John the Evangelist's, London (d. 1451).

Powicke, *Medieval Books*, no. 398.

691. **St. John's College 207** Saec. XIII$\frac{3}{4}$
Bible (140 × 90)

Good miniature, historiated initials. Good penwork initials [Illus.]

Prov. Given in 1610 by William Laud, president, archbp. of Canterbury 1633–41, fol. 1, 1^{v}.

Exh., *Archbishop Laud*, no. 1.

692. **Lincoln College Lat. 150** Saec. XIII $\frac{3}{4}$ Paris

Missal (fragment) (185 × 130)

Fine miniatures, historiated initials. Fine penwork initials. Close in style to the Franciscan Missal (Assisi, Museo del Sacro Convento) of the Ste Chapelle Group. [Illus.]

Prov. Premonstratensian abbey of Dommartin, Pas de Calais, 'A M. Oblin, abbé de Dommartin' (d. 1824), fol. 1. Sir Thomas Phillips MS. 4446, fol. 1, 1830. Chester Beatty Collection, 1925. Sale, Sotheby's 24 June, 1969, lot 52, pl. 18. Given in 1975 by Mr. Eric Sexton, Camden, Maine, U.S.A.

E. G. Millar, *The Library of A. Chester Beatty*, II, 1930, 63–5, pl. CXXI a, b; Branner, 128 n. 30, cat. 236.

693. **Queen's College 299** *c.* 1280–90 Paris

Bible (320 × 210)

Fine historiated and other initials. Good penwork borders, initials. By the workshop of the later Ste Chapelle Evangeliary, Brit. Libr. MS. Add. 17341. [Illus.]

Prov. Given by Griffith Powell, fellow of Jesus College, 1599, fol. 1.

694. **Lincoln College Lat. 24** *c.* 1280–90 Paris

Bible (250 × 175)

Fine historiated and other initials. Good penwork initials. Close in style to the later Ste Chapelle Evangeliary, Brit. Libr. MS. Add. 17341. [Illus.]

Prov. Given by Robert Flemmyng, dean of Lincoln (d. 1483), fols. 402^{v}, 403.

Emden, 700.

695. **New College 3–6** *c.* 1290 Troyes

Bible (480–98 × 320–55)

Fine arabesque and penwork initials. Marginal drawings by the scribe. Faint marginal sketches, MS. 3, fols. 155–6, 159, 170; MS. 6, fol. 289. Written by Philippus presbiter of Troyes (colophon MS. 3, fol. 435^{v}; MS. 4, fol. 411). He also wrote a Missal of Paris use, Bibl. St. Geneviève MS. 98, and a fragment (Guillaume de Bourges, Office) inserted into Troyes, Bibl. Mun. MS. 1148, fols. 282–90, where a note of saec. XVIII identifies him with a monk at Moutier-la-Celle, near Troyes.

Prov. Given by Thomas Bekyngton, bp. of Bath and Wells (d. 1465).

'Notable accessions', *B.L.R.* VI (1960), 571; Hunt, *Medieval Library*, 326.

696. **Balliol College 43** Saec. XIII ex. Paris (?)

Thomas Aquinas, Summa theologiae, secunda pars, pt. II (350 × 234)

Good border, initial. Penwork initials.

Prov. Bequeathed by magister William Burnell (d. 1304), fol. 1^{v} (erased inscription). Bequeathed by William Duffeld, fellow of Merton College (d. 1453), at the request of magister William Moreton, fellow of Balliol College, fol. 1.

Emden, 602; Mynors, *Catalogue*, 31.

697. **Corpus Christi College 111** Saec. XIII ex.

Aristotle, Corpus Vetustius (224 × 180)

Good and other initials (mutilated), Penwork initials, scribbles. Pen drawing, saec. XIV (?), fol. 23.

Prov. Dr. John Dee (d. 1608), fol. 1. Bequeathed by Brian Twyne, fellow (d. 1644), fol. 1.

James, *Dr. John Dee*, 30; *Aristoteles Latinus*, I, 407; D. Callus, 'Aristotelian learning in Oxford', *Proceedings of the British Academy*, XXIX (1943), 237 n., 268.

698. **All Souls College 95** Saec. XIII ex.

Ptolemy, Almagest (390 × 250)

Good penwork borders, initials, diagrams.

Medieval binding of white leather on boards, saec. XV/XVI.

699. **All Souls College 71** Saec. XIII ex. South France (?)

Medical treatises (310 × 205)

Fine historiated and other initials. Penwork initials [Illus.]

700. **Balliol College 2,** fols. 132^{v}–532 Saec. XIII ex.

Bible (270 × 195)

Fine historiated and other initials in two styles, by a French illuminator perhaps working in Italy, cf. no. 895. Preliminary marginal drawings. Good penwork initials. Note concerning payment for these, fol. 177^{v}. [Illus.]

Prov. Sold by Hugo Lynke, executor of Nicholas Josseham (?), to George Hyll, cleric, in 1486, erased note fol. 532^{v}. Thomas Grace, fol. 181, saec. XVI.

Mynors, *Catalogue*, 2.

701. **Merton College 270** (O.3.2) Saec. XIII ex.

Aristotle, de animalibus. Ps.-Aristotle, Problemata (335 × 248)

Fine border, initial. Good penwork borders, initials. Marginal drawings, fols. 109^{v}, 124, saec. XIV.

Prov. John of Osmyngton, fellow, rector of Fuggleston, Wilts, in 1333, fol. i^{v}.

Emden, 1408.

702. **Christ Church 98** Saec. XIII ex. North France

Psalter (234 × 170)

Fine borders, historiated and other initials. [Illus.]

Prov. Obit of Eleanor, fol. 7, saec. XIV (?). Notes and obits of the Rede and Cottismore families. 1430–66, passim.

703. **All Souls College 79** Saec. XIII ex.

Gilbertus Anglicus, Compendium medicinae (412 × 280)

Fine borders, historiated initials. Penwork borders, initials. [Illus.]

Medieval boards, under faded pink leather, rebacked.

Prov. Sold by the executors of John Fodyngton to Elias Stokes, fellow of Exeter College, rector 1391–3 (d. 1436), fol. 166^{v}.

Emden, 1780.

704. **All Souls College 27** Saec. XIII ex.

Thomas Aquinas, Summa theologiae, secunda pars, pt. I (275 × 250)

Fine border (mutilated), historiated initial. Good penwork borders, initials. 'Deo gratias, Dionisius (scribe ?) sit benedictus. Amen', fol. 177. [Illus.]

Medieval binding of white leather on boards, rebacked.

705. **Exeter College 69** Saec. XIII ex. French (?)

Philippus Gualterius, Poems (192 × 130)

Penwork initials. Rough marginal drawings. Written by Petrus heremita, 1288, fol. 80^{v}.

706. **Balliol College 103** Saec. XIII ex.

Albertus Magnus, de animalibus (330 × 225)

Good initials. Penwork initials.

Prov. Bequeathed by William Saxton, fellow (d. 1424), who pledged it jointly with Robert Saxton in a loan chest in 1415.

Emden, 1648–9; Mynors, *Catalogue*, xix.

707. **New College 318** Saec. XIII^{2}

Bible (incomplete) (258 × 175)

Good historiated and other initials (damaged and mutilated).

708. **Keble College 26** Saec. XIII^{2} Paris

Peter Lombard, Sentences with gloss (324 × 213)

Good initials. Possibly by the same atelier as MS. Keble College 20. Cf. no. 678.

Prov. Bequeathed by Dr. James E. Millard (d. 1894).

M. B. Parkes, 'The influence of the concepts of *Ordinatio* and *Compilatio* on the development of the book', *Medieval Learning and Literature. Essays presented to R. W. Hunt, eds.* J. J. G. Alexander, M. T. Gibson, 1976, 126–7, 139, pl. XIII.

709. **All Souls College 74** Saec. XIII^{2} South France

Isaac Judaeus, Liber dietarum, etc. (252 × 170)

Good initials. Penwork initials.

710. **Queen's College 70** Saec. XIII^{2} French (?)

Bible (245 × 185)

Fine historiated initials. Good penwork borders, initials. [Illus.]

711. **New College 22** Saec. XIII^{2}

Isaiah and Jeremiah with gloss (mutilated) (357 × 245)

One fine historiated initial, other good initials.

712. **Merton College 76** (I.1.5) *c.* 1300 French (?)

Thomas Aquinas, Summa theologiae, prima pars (325 × 225)

Fine border, historiated initial. Good penwork initials. [Illus.]

Prov. Bequeathed by Thomas de Staundon, fellow 1352–3, fol. 3^{v}.

Emden, 1757.

713. **Merton College 75** (I.1.4) *c.* 1300 French (?)

Thomas Aquinas, Summa theologiae, tertia pars. (320 × 218)

Good initial. Good penwork initials.

Prov. Possibly bequeathed by Thomas de Staundon, fellow 1352–3.

Powicke, *Medieval Books*, no. 218; Emden, 1757.

714. **University College 129** *c.* 1300

Johannes de Balbus, Catholicon (402 × 280)

Fine border, initials, added to an Italian MS. with penwork initials, saec. XIII^{2}. Mutilated.

Medieval binding of white leather on boards.

Prov. Given by John de Pontissara (Pontoise), bp. of Winchester (d. 1304), to the Chapel of the Hospital of St. Elizabeth of Hungary, Winchester, fols. i, 378^{v}.

William Shippen, fellow, rector of Stockport, Cheshire (d. 1693), fol. i.

Hunt, *Manuscript*, 22 n. 4.

715. **All Souls College 72** *c.* 1300 South France

Nicolaus, Antidotarium. John Damascene, Aphorisms, etc. (335 × 200)

Good borders, historiated and other initials. By two hands. Penwork borders, initials. [Illus.]

Medieval boards.

Prov. Calendar of Maguelonne, near Montpellier, fols. iv–vi. Pledged in the Rothbury chest, Oxford, in 1363, fol. 294.

716. **Merton College 123–6** (M.2.7–10) Saec. XIV in. Paris

Vincent of Beauvais, Speculum historiale (370 × 245)

Fine borders, historiated and other initials. Good penwork borders, initials. Marginal sketch of border ornament, vol. IV, fol. 3^v. Illumination by two hands. Vol. II written by Robert of Longjumeau, fol. 280. Vol. III written by John Wade, fol. 287^v, an English scribe working in Paris where he wrote Reims, Bibl. Mun. MS. 734. He also wrote Toulouse, Bibl. Mun. MS. 449. Corrector's marks at the end of vols. I and II. [Illus.]

Prov. Bequeathed by Roger Martival, chancellor of Oxford University 1293–5, bp. of Salisbury (d. 1330), Edmund Palmer, fellow (d. 1543), vol. IV, fol. 290^v.

Powicke, *Medieval Books*, no. 143.

717. **Magdalen College Lat. 187** Saec. XIV in.

Aristotle, Logica. Gilbert de la Porrée, de sex principiis, etc. (320 × 215)

Fine borders, initials. Good penwork initials.

Prov. Deposited in a loan chest by Nicholas Serle, rector of Treeton, Yorks., 1 Apr., 1441, and by M. Burley, 1486, fol. i.

Emden, 1670.

718. **Balliol College 275** Saec. XIV in. South France (?)

Castor, Epistola ad Cassianum. Cassian, Collationes (340 × 240)

Fine border, historiated initial. Fine penwork initials. [Illus.]

Prov. William Gray, bp. of Ely 1454–78, fol. 1^v.

719. **Lincoln College Lat. 109,** fols. 18–105 Saec. XIV in.

Aegidius de Columna, Comm. on Sentences, Bk. I (395 × 270)

Good border, historiated initial. Good penwork initials. [Illus.]

Medieval binding of white leather on boards.

Prov. Pledged in the Winton chest, Oxford, 9 Jan. 1427, by John Madderby, fellow, rector of St. Ebbe's, Oxford, fol. i. Bought by Thomas Wyche, canon of Lichfield (d. 1475), for xvi s. from John More, the stationer, 1 Aug. 1440.

Emden, 1202, 2102.

720. **Balliol College 231** Saec. $XIV\frac{1}{4}$ Paris (?)

Galen, Various works (355 × 252)

Good border, initial. Good penwork borders, initials (italianate).

Prov. Bequeathed by Simon de Holbeche, M.D., fellow of Peterhouse, Cambridge (d. 1334), in memory of Stephen of Cornwall, master of Balliol College in 1307, fol. 1^v.

Emden, 491, 945; Ker, *College Libraries*, 298 n. 27, 299 n. 28, 300 n. 35.

721. **Balliol College 232A** Saec. $XIV\frac{1}{4}$ Paris (?)

Aristotle, Physics (405 × 273)

Fine borders, fine and other initials (mutilated). Good penwork border, initials. [Illus.]

Prov. Thomas Tyrwyte, fellow, master of Balliol College (d. by 1399), fols. 103^v, 104, 104^v. William Gray, bp. of Ely 1454–78, fol. ii^v.

Emden, 1925; Ker, *College Libraries*, 298 n. 27, 300 n. 35.

722. **Corpus Christi College 65** Saec. $XIV\frac{1}{4}$ South France (?)

Tacuinus, de cura morborum (435 × 285)

Fine border, initial. Fine penwork initials.

Prov. Ex libris of St. Augustine's, Canterbury, saec. XIV, fol. 1. Pressmark 'D.14 G.2', saec. XV ex., fol. 1. Notes of accounts possibly by Dr. John Dee (d. 1608). Bequeathed by Brian Twyne, fellow (d. 1644), fol. i^v.

James, *Ancient Libraries*, 335, no. 1197; *id.*, *Dr. John Dee*, 36.

723. **All Souls College 21** Saec. $XIV\frac{1}{4}$

Thomas Aquinas, Summa theologiae, prima pars (345 × 230)

Fine border, historiated initial. Good penwork borders, initials. Made for a patron, possibly a Cistercian monk, adoring Christ crucified on fol. 5. [Illus.]

Prov. Ex libris of Vaudey abbey, Lincs., fol. 177^v. Narcissus Luttrell, 1677, fol. ii.

724. **Merton College 44** (C.1.5), fols. 166–177^{v} Saec. xiv$\frac{1}{4}$

Elucidarium (245 × 175)

Good historiated initial. Penwork initials. [Illus.]

725. **St. John's College 112,** fols. 1–114 Saec. xiv$\frac{1}{4}$

Alphabetum narrationum (242 × 158)

Border, initial. Written by Arnuldus de Serain (cf. Kaeppeli). '. . . 1308', fol. 114.

Prov. Cf. no. 143.

T. Kaeppeli, *Scriptores Ordinis Praedicatorum Medii Aevi*, 1970, i. 132.

726. **St. John's College 204** Saec. xiv$\frac{2}{4}$ North France

Psalter (mutilated). In Latin and French (147 × 95)

Fine borders, historiated and other initials. Good penwork initials, line-fillers. [Illus.]

Prov. Calendar suggests French Flanders. Robert Travers, commoner, 1607, fol. 1.

727. **Balliol College 302** Saec. xiv$\frac{2}{4}$

Duns Scotus, Ordinatio I (380 × 265)

Fine borders, initials (mutilated). Good penwork borders, initials. By the illuminator of MS. Balliol College 208, originally bound up with it. Partly written by the same scribe. Cf. no. 728.

728. **Balliol College 208** Saec. xiv$\frac{2}{4}$

Willelmus de Alnwick. Duns Scotus, Ordinationes ii, iii, etc. (380 × 246)

Fine borders, initials. Good penwork borders, initials. By the same illuminator as MS. Balliol College 302, originally bound up with it, cf. list of contents, fol. 184^{v}. Partly written by the same scribe. Cf. no. 727.

Prov. Fr. Andrea and fr. Jacobus de Tholomeis (later bp. of Narni, d. 1390 (?)), fol. 184^{v}.

Ker, *College Libraries*, 298 n. 27, 299 n. 28, 300 n. 35.

729. **Merton College 100** (I.3.5) Saec. xiv$\frac{2}{4}$ Paris (?)

William Ockham, Comm. on the Sentences (320 × 250)

Fine penwork borders. Fine and other penwork initials.

730. **Merton College 99** (I.3.4) Saec. xiv$\frac{2}{4}$

Thomas Aquinas, Comm. on the Sentences, Bk. IV (350 × 235)

Fine border, historiated initial. Good penwork initials [Illus.]

731. **St. John's College 98** Saec. xiv^{1}

Misc. religious pieces (250 × 180)

Good penwork initials. Marginal sketches, fols. 111, 214^{v}.

Prov. Unidentified shelfmark 'Liber tercii loci in quarta stacione', saec. xiv, fol. 1. Given in 1618 by William Laud, president, archbp. of Canterbury 1633–41, fol. 1.

Exh., *Archbishop Laud*, no. 6.

732. **Oriel College 44** Saec. xiv^{1} French (?)

Bible concordances (305 × 210)

Initial.

733. **Merton College 227** (N.1.1) Saec. xiv^{1} South France

Razes, Almansor, etc. Avicenna, de animalibus (295 × 204)

Good historiated initial. Penwork initials. [Illus.]

Prov. Given by magister Nicholas Sandwich to William Reed, fellow, bp. of Chichester (d. 1385), who bequeathed it to the College, fol. ii^{v}.

Emden, 1558.

734. **Merton College 149** (M.1.1) Saec. xiv med.

Gerard of Bologna, Summa theologiae (282 × 200)

Good border, initial. Penwork initials.

Prov. Bequeathed by William Reed, fellow, bp. of Chichester (d. 1385), who acquired it from Thomas Trillek, bp. of Rochester (d. 1372), fol. 1^{v}.

Emden, 1558.

735. **Lincoln College Lat. 50** Saec. xiv med.

Durandus, Rationale divinorum officiorum (370 × 260)

Good penwork initials.

Prov. Notes by Thomas Gascoigne, chancellor of Oxford University (d. 1458), fol. 1. Given in 1436 by John Southam, archdeacon of Oxford, fol. i^{v}.

Weiss, *Catalogues*, 348, 354–5; Emden, 746, 1733.

736. **Lincoln College Lat. 1** Saec. xiv med.

Thomas Aquinas, Comm. on the Sentences, Bk. IV (305 × 240)

Fine border, initial (mutilated). Fine penwork borders, initials.

Prov. Given by Robert Flemmyng, dean of Lincoln (d. 1483), fol. 1.

Weiss, *Catalogue*, 346; Emden, 700.

737. **All Souls College 51** Saec. xiv med.

Justinian, Infortiatum with gloss of Accursius (455 × 290)

Fine miniature, borders, fine and other initials. Good penwork initials, line-fillers, catchwords. Fine borders, initials in a somewhat later style perhaps added, fols. 1,

10. 'Vinum scriptori det Petrus de Meliori, nomen scriptoris Guillelmus Plenus-amoris', colophon, fol. 279. [Illus.]

Binding of white leather on boards, saec. xv/xvi, (?).

Prov. Defaced, fol. 1.

738. **St. John's College 164** *c.* 1364–73 Paris

Astronomical treatises, etc. including Pèlerin de Prusse. In French (223 × 140)

Fine miniatures, borders. Fine historiated and other initials (some defaced), diagrams. Horoscopes of Charles V and his family added after 1373, fols. 158^{v}–161^{v}. [Illus.]

Prov. Pèlerin de Prusse dedicated his treatise to Charles V as Dauphin, *c.* 1361–2. The miniatures show Charles as king, however. At Vincennes in 1418 (cf. Delisle). Sir William Paddy, commoner, physician to James I, 1633, fol. iv^{v}.

L. Delisle, *Recherches sur la librairie de Charles V*, I, 1907, 266, no. 182; L. Thorndike, *A history of magic and experimental science*, III, 1934, 586–7, 589 n. 10; Exh., *La Librairie de Charles V*, Bibliothèque Nationale, Paris, 1968, no. 199, pl. 5; C. R. Sherman, *The portraits of Charles V of France*, 1969, 23, 31, *pls. 6, 70*; C. R. Sherman, 'Representations of Charles V of France as a wise ruler', *Medievalia et Humanistica*, 2 (1971), 87, figs 4 & 5;Exh., *Les Fastes du Gothique. Le siècle de Charles V*, Paris, 1981, no. 289.

739. **Merton College 298** (O.2.8) Saec. xiv$\frac{3}{4}$

Pierre Bersuire, Repertorium morale (395 × 285)

Fine borders, fine and other initials (mutilated).

Prov. Given in 1466 by Henry Sever, warden (d. 1471), fol. ii^{v}.

Emden, 1673.

740–1 *Vacant*

742. **University College 7** Saec. xiv$\frac{3}{4}$

Psalter (90 × 57)

Fine borders, initials. Fine penwork initials.

Prov. Note (largely erased), fol. 138^{v}, saec. xv ex., 'Jehan . . . (diepp . . . enfant de . . . cathedralle) . . . de Rouen'.

Frere, I, 475.

743. **Balliol College 64** Saec. xiv^{2}

Petrus de Candia, Comm. on the Sentences (incomplete) (330 × 230)

Fine borders, initials. Penwork initials.

Prov. Johannes Berverius, B.C.L. (Paris), 1379, fol. 150; Guillelmus Pulcrinepos (Guillaume Beauneveu (?), B.D. Paris, 1403), fol. i. In Italy, fol. 1, saec. xv^{1}. William Gray, bp. of Ely 1454–78, fol. ii^{v}.

744. **Keble College 15** Saec. xiv ex. Avignon

Book of Hours. Use of Rome. In Latin and French. (192 × 140)

Fine miniatures, borders. Fine historiated and other initials. By the same hand as Paris, Bibl. Nat. Lat. 10527. Parchment of Italian type. [Illus.]

Contemporary boards with stamped leather cover, resewn, saec. xv^{2}, North French (?). Gold tooling added to four corners, saec. xix.

Prov. J. A. J. Delignières de Bommy d'Abbeville, first pastedown, fol. 1. Bequeathed by Dr. Henry P. Liddon, canon of St. Paul's, London (d. 1890).

745. **University College 9** Saec. xiv ex.

Breviary. Carmelite use (195 × 125)

Fine borders, initials. Fine penwork initials.

Blind-panels stamped leather binding on boards, English, *c.* 1530.

Prov. Made for English use. More English saints added in the calendar, saec. xv. Obits 'matris mee', 1585. Ann Mansfield, Ann Conyers, saec. xvi.

746. **New College 197–198** Saec. xiv ex. Paris

Anonymous, Quaestiones on Decretals, Bks. III–V (395 × 265)

Fine borders, historiated and other initials (mutilated). Fine penwork initials. [Illus.]

Prov. Given by Richard Andrew, fellow, dean of York (d. 1477), fol. 1.

747. **Balliol College 56** Saec. xiv ex. Paris (?)

Franciscus de Perusio, Quaestiones on Sentences (incomplete) (290 × 205)

Fine borders, historiated initials (mutilated). Good penwork initials. [Illus.]

Prov. William Gray, bp. of Ely 1454–78, fol. ii^{v}.

748. **Keble College 11** *c.* 1400 North-west France

Book of Hours. Use of Rome (170 × 118)

Fine miniatures, borders, initials. [Illus.]

Prov. Calendar suggests the diocese of Nantes. 'Questo libro e di Signore. . . .', fol. i. Bequeathed by Dr. Henry P. Liddon, canon of St. Paul's, London (d. 1890).

749. **All Souls College 89** *c.* 1400
Robert de Eumodio, Cato moralizatus (incomplete) (368 × 265)

Good borders, initials. Good penwork initials.

Prov. Given by Richard Andrew, warden 1437–42, dean of York (d. 1477), fol. 2.

750. **New College 189–190** *c.* 1400
Paris (?)
Henri de Bohic, Comm. on Gregory IX, Decretals (430 × 285)

Fine borders, historiated and other initials. Fine penwork initials. Signed 'H. Bolaye', MS. 189, fol. 143 (scribe ?). [Illus.]

Prov. Given by Thomas Bekyngton, fellow, bp. of Bath and Wells 1443–65, fol. 1.

Hunt, *Medieval library*, 326 n. 29.

751. **New College 154** *c.* 1400
South France
Philippus de Pergamo, Comm. on Cato moralizatus (330 × 235)

Fine borders, fine and other initials. [Illus.]

752. **All Souls College 57** Saec. XV in.
Cino da Pistoia, Comm. on Justinian, Codex (425 × 265)

Fine border (mutilated), fine penwork borders, initials (unfinished).

Binding of white leather on boards, saec. XV/XVI.

Prov. Given by Richard Andrew, warden 1437–42, dean of York (d. 1477).

Emden, 35.

753. **Magdalen College Lat. 41** Saec. XV in.
Bernard, Augustine, etc. In French (260 × 170)

Fine penwork borders, initials, line-fillers.

Prov. Given to Barking Abbey by Elizabeth de Vere *c.* 1477, fol. 218^v. Given by Sir Arthur Throckmorton, 1626, fol. 1.

A. J. Doyle, 'Books connected with the de Vere family', *Essex Archaeological Society Transactions*, XXV new ser. (1958), 235 ff.

754. **All Souls College 59** 1406
Orléans
Johannes Faber, Comm. on Justinian, Institutiones (412 × 270)

Fine borders (mutilated), fine and other initials. Written by Johannes Legam at Orléans, 1406, fol. 245.

Binding of white leather on boards, saec. XV/XVI (?).

Prov. James Goldwell, fellow, bp. of Norwich 1472–99, first pastedown.

N. R. Ker, *Records of All Souls College Library 1437–1600*, 1971, 131.

755. **New College 185–6** Saec. XV$\frac{1}{4}$
Paris
Henri de Bohic, Comm. on Gregory IX, Decretals. Bks. I–V (447/425 × 300)

Fine borders (mutilated), fine and other initials. Fine penwork initials. Marginal drawings, (catchwords, explicit). Completed in 1417, MS. 186, fol. 342^v. [Illus.]

Prov. Given in 1503 by William Spekyngton, fellow, to William Warham, archbp. of Canterbury 1503–32, who bequeathed it to the College.

Emden, 1741, 1990; Hunt, *Medieval library*, 329–30.

756. **Keble College 39** Saec. XV$\frac{1}{4}$
Paris
Book of Hours. Use of Paris. In French (210 × 158)

Important miniatures. Fine historiated and other borders, historiated and other initials. Line-fillers. By the Master of the Duke of Bedford and his associates. [Illus.]

Prov. Calendar suggests Paris (Ste. Geneviève in gold). 'Janvier 1834', erased, fol. iii. William Tite, July 1857, fol. i^v. His sale, Sotheby, Wilkinson and Hodge's, 18 May 1874, lot 1486. Sir Thomas Brooke, Huddersfield (d. 1908). Bequeathed by his brother, Canon C. E. Brooke (d. 1911).

757. **Lincoln College Lat. 14** 1415
Savoy (?)
Nicholas Bonet, Metaphysics. Augustine, de cognitione verae viae (215 × 140)

Good and other initials, marginal sketch, fol. 52^v. Written in 1415, fol. 98.

Prov. Anthon Bothun 'de Sabaudia . . .', 1423, fol. 1. Bought in Savoy by John Tregarran from Guillelmus Secomps, a bookseller, 1424, fol. 1. Request for prayers for Thomas Gascoigne, chancellor of Oxford University (d. 1458) and John Tregarren, fol. 106. Given by them jointly to the College, fol. 5^v.

Weiss, *Catalogues*, 350, (90) n. 114; W. A. Pronger, 'Thomas Gascoigne', *English Historical Review*, LIII (1938), 618; Emden, 747, 1893.

758. **Magdalen College Lat. 5** *c.* 1420–30
North France
Book of Hours (incomplete). Unidentified use (220 × 160)

Fine miniatures, borders, initials, line-fillers. [Illus.]

759. **Keble College 16** Saec. xv^2_4
Book of Hours. Use of Paris. In Latin and French (193 × 137)

Fine miniatures, borders. Fine and other initials. [Illus.]

Original boards with blind-stamped leather Annunciation panel in the centre of the top cover, French, saec. xv.

Prov. Given to Charlotte Jacquot in 1557 by 'mons Jehan Jacquot' conseiller du Roy et maitre ordinaire' of Dijon, fols. i, 173. Family chronicle for 1566–8 added by Estienne Deloisye of Dijon who married Charlotte Jacquot in 1566, fols. 173–4. Bequeathed by Dr. Henry P. Liddon, canon of St. Paul's, London (d. 1890).

760. **Keble College 38,** fols. 1–436^v and fols. 440–95^v. Saec. xv^2_4
Missal. Dominican use. In French. Book of Hours. Use of Paris (158 × 110)

Important miniatures, borders. Fine historiated and other initials. Line-fillers. By the Master of sir John Fastolf (cf. Pächt, Alexander, I, nos. 670, 695, 696). Cf. no. 763. [Illus.]

Prov. Originally two separate books, made in the same workshop for a lay patron, a tertiary of the Dominican Order who may have been a member of the French royal court. Colophon with a coat of arms, fol. iv^v, a faked provenance. William Bragge sale, Sotheby, Wilkinson and Hodge's, 7 June 1876, lot 356. Sir Thomas Brooke, Huddersfield (d. 1908). Bequeathed by his brother, Canon C. E. Brooke (d. 1911).

J. J. G. Alexander, 'A Lost Leaf from a Bodleian Book of Hours', *B.L.R.* VIII (1971), 250; L. M. J. Delaissé. 'A liturgical problem at the end of the Middle Ages', *Essays presented to G. I. Lieftinck. Litterae textuales. IV, eds.* J. P. Gumbert, M. J. M. de Haan, 1976, 16–27, figs. 4, 5.

761. **St. John's College 102** Saec. xv^1
Roman de Melibée et Prudence, etc. In French (255 × 155)

Good borders, initials (defaced).

Prov. Christopher Coles, commoner, 1611, fol. 3.

762. **St. John's College 197** 1450 Avignon
Guido, Comm. on Antidotarium; misc. medical texts (190 × 103)

Fine borders, fine and other initials. Fine penwork initials. Written by Jo. Durandus in Avignon in 1450, fol. 124^v. Fine borders in the same style added, fols. 291–372, to an Italian MS. of saec. XIII^2_4. [Illus.]

Prov. Sir William Paddy, commoner, physician to James I, 1634, fol. 3.

763. **St. John's College 208** Saec. xv med.
Offices and prayers. In Latin and English (130 × 80)

Fine miniatures, borders, historiated and other initials. By the Master of sir John Fastolf. Cf. no. 760. [Illus.]

Prov. Arms of Baron, fol. 89^v. Henry Warner, 1636, fol. 1.

J. J. G. Alexander, 'A Lost Leaf from a Bodleian Book of Hours', *B.L.R.* XIII (1971), 251 n. 6.

764. **Wadham College 10** (A.10.39) Saec. xv med.
Book of Hours (mutilated). Use of Evreux. (150 × 110)

Good miniatures (damaged, some retouched), borders. Fine and other initials. [Illus.]

Prov. Given in 1720 by William Shortgrave, scholar of Wadham College in 1692, fol. 2.

765. **Regent's Park College 10.G** Saec. xv med. North France or French Flanders
Book of Hours (incomplete). Use of Rome. In Latin and French (218 × 156)

Good miniatures (some defaced), borders, good and other initials. Line-fillers. [Illus.]

Binding of blind-stamped leather, re-backed, Flemish, saec. xv^2.

Prov. Given by the executors of Dr. Horsepool, 1885.

766. **Merton College 303** (N.1.5) Saec. xv med.
Isidore, Sententiae, Etymologiae, etc. (304 × 214)

Good and other penwork initials. Diagrams.

767. **Queen's College 405** Saec. xv^3_4
Book of Hours. Use of Bourges. In Latin and French (184 × 143)

Good miniatures. Good historiated and other initials, line-fillers. [Illus.]

Prov. Calendar points to Bourges, Book-plate (partly torn) with arms of Henrich Freijher . . . Hohenburg . . . 1674, first pastedown. Hans Heinrich von Howart Eninische Kanzler, 1697, last pastedown. Book-plate with arms of William Jackson, archdeacon of Carlisle, provost (d. 1878), fol. 1. Given by Georgiana his daughter, wife of J. R. Magrath, provost (d. 1930).

768. **Keble College 6** Saec. xv^3_4 French Flanders
Book of Hours. Use of Rome (113 × 80)

Good borders, historiated and other initials. Penwork initials.

Prov. Calendar suggests diocese of Cambrai. Pierre Martin Beaupère, 1726, fols. iii^v, 183^v. Bequeathed by Dr. Henry P. Liddon, canon of St. Paul's, London (d. 1890).

769. **Keble College 56** Saec. xv$\frac{3}{4}$ Rouen

Missal. Use of Rouen. In Latin and French (274 × 185)

Fine miniatures. Historiated and other borders. Fine historiated and other initials. Principal illustrations by the same artist as Aix-en-Provence, Bibl. Mun. MS. 22, called by F. de Mély, 'T. Chugoinot' (cf. *Les Primitifs et leur signatures*, 1913, 185 ff., pl. IX). [Illus.]

Prov. Written for an Augustinian canon shown kneeling in the borders. William Bragge sale, Sotheby, Wilkinson and Hodge's, 7 June 1876, lot 364. Sir Thomas Brooke, Huddersfield (d. 1908). Bequeathed by his brother Canon C. E. Brooke (d. 1911).

770. **Lincoln College Lat. 152** Saec. xv$\frac{3}{4}$

Book of Hours. Use of Paris (143 × 100)

Fine miniatures, borders, initials. By Maître François (cf. no. 777). [Illus.]

Prov. Given *c.* 1972–73 by Mr. Eric Sexton, Camden, Maine, U.S.A.

771. **St. John's College 117** Saec. xv$\frac{3}{4}$

Terence, Comedies (235 × 156)

Fine borders. Fine and other initials.

Prov. '1477', 'Quod Hanley', fol. 1. Given in 1610 by William Laud, president, archbp. of Canterbury 1633–41, fol. 2.

Exh., *Archbishop Laud*, no. 2.

772. **Merton College 300** (A.3.1), fols. 157–207 Saec. xv$\frac{3}{4}$

Seneca, Tragedies (423 × 270)

Border (mutilated), fine initials. Written in semi-humanistic script by Florius, a Florentine scribe working in France *c.* 1449–80.

Prov. Cf. no. 900.

C. F. Finlayson, 'Florius Infortunatus', *Scriptorium*, XVI (1962), 379 n. 10.

773. **Keble College 57** Saec. xv$\frac{3}{4}$ Valence (Drôme)

Missal. Use of Valence (330 × 225)

Good miniatures, borders, on inserted bifolium, fols. 141–2. Fine borders, initials, by a different hand. Cadels. [Illus.]

Contemporary binding of blind-stamped leather on boards, probably Valence.

Prov. Calendar of Valence cathedral. Arms of de Sassenage, fol. 141^{v}, and of Commiers, fol. 142, both families from Dauphiné. Note on the marriage of Laurens de Sassenage to Marguerite Dorayson, 1557, fol. 2^{v}. Sir Thomas Brooke, Huddersfield (d. 1908). Bequeathed by his brother Canon C. E. Brooke (d. 1911).

774. **Keble College 78** Saec. xv$\frac{3}{4}$

Book of Hours. Use of Paris (156 × 107)

Good miniatures (retouched), borders. Good and other initials. Line-fillers.

Prov. Liturgical evidence for Paris. Family chronicle for 1476–98 added by Anthoine Carrel, fols. 145–7. Sold by Joseph T. Hand in 1841 to William Vines who gave it to William H. Black, F.S.A. (d. 1872). Rev. Christie Chetwynd Atkinson (d. 1911).

775. **Keble College 40** *c.* 1470–80 Lyons

Book of Hours. Use of Rome (139 × 94)

Fine miniatures. Fine and other initials. Line-fillers. Penwork initials with calligraphic flourishes. [Illus.]

Prov. Calendar suggests the region of Lyons (St. Nicetius in red). Sir Thomas Brooke, Huddersfield (d. 1908). Bequeathed by his brother, Canon C. E. Brooke (d. 1911).

776. **Keble College 29** 1473 Dijon

Breviary, Carthusian use (150 × 108)

Fine miniatures (partly rubbed), borders. Fine and other initials. Written by Johannes Radulphus Rouvre for Fr. Hugo Grignardi of the Charterhouse of the Holy Trinity at Champmol, nr. Dijon, fol. 418^{v}.

Original boards and sewing with brown leather cover, metal plaque on the back with monogram 'M.A.'.

Prov. Ex libris of the Charterhouse at Champmol, nr. Dijon, fol. 588^{v}, and below, 'datus est patri philo Guignaut a patribus Carthusiae', remains of a seal. Sir Thomas Brooke, Huddersfield (d. 1908). Bequeathed by his brother, Canon C. E. Brooke (d. 1911).

L. Delisle, 'Bibliographie', *Bibliothèque de l'École des Chartes*, LIII (1892), 184, no. IX.

777. **Christ Church 93** After 1484

Hours, etc. of St. Denis (195 × 125)

Fine miniatures, borders, initials. By Maître François (cf. no. 770). [Illus.]

Prov. Arms of Guy XV, comte de Laval Montmorency (d. 1500), with the collar of the Order of Saint-Michel to which he was probably admitted in 1484. He is shown kneeling in prayer, fol. 1.

778. **Keble College 7** Saec. XV ex.

Book of Hours. Use of Rome (151 × 90)

Good miniatures. Initials, line-fillers. Cadels [Illus.]

Prov. Arms added, fol. 21^{v}: quarterly, 1 and 4 checky of nine or and gules nine knots (?) gules and or, 2 and 3 gules a chief argent. 'Ex libris, Htibordes', first pastedown Bequeathed by Dr. Henry P. Liddon, canon of St. Paul's, London (d. 1890).

779. **Keble College 76** Saec. XV ex.
Books of Hours. Use of Langres (197 × 138)

Good miniatures, borders, historiated and other initials. Line-fillers.

Prov. Liturgical evidence suggests use in the diocese of Langres. Rev. Christie Chetwynd Atkinson (d. 1911).

780. **Keble College 42** Saec. XV ex.
Book of Hours (incomplete). Use of Rome (152 × 100)

Fine miniatures, historiated and other borders. Fine and other initials. By several hands. The principal miniatures by Jean Colombe. First leaf, fol. 1–1^{v}, is a restoration by Caleb Wing, *c.* 1850. [Illus.]

Prov. John Boykett Jarman sale, Sotheby, Wilkinson and Hodge's, 13 June 1864, lot 49. Sir Thomas Brooke, Huddersfield (d. 1908). Bequeathed by his brother, Canon C. E. Brooke (d. 1911).

781. **Keble College 41** Saec. XV ex.
Book of Hours. Use of Rome (127 × 90)

Fine miniatures, borders. Good and other initials. Line-fillers. From the atelier of Jean Colombe. [Illus.]

Prov. Made for a lady, member of the Menon family from Dauphiné, represented with her arms on fol. 34^{v}. Sir Thomas Brooke, Huddersfield (d. 1908). Bequeathed by his brother Canon C. E. Brooke (d. 1911).

782. **Keble College 9** Saec. XV ex.
Book of Hours. Use of Troyes. In Latin and French (184 × 130)

Fine miniatures, borders, initials, line-fillers.

Prov. Calendar of Troyes. Lallemant, 1758, fol. 1. Acquired from F. W. Kerslake, 1875, by Dr. Henry P. Liddon, canon of St. Paul's, London, who bequeathed it to the College (d. 1890), *ex libris* first pastedown, fol. i^{v}.

Frere, I, 163 n.

783. **Campion Hall 2** Saec. XV ex.
Book of Hours. Use of Rome (?) (192 × 125)

Good miniatures, borders, initials.

Prov. Given by sir Frank Brangwyn, R.A. (d. 1956).

784. **Lincoln College Lat. 120** Saec. XV ex.
Book of Hours. Use of Sens (183 × 123)

Good borders, initials.

Prov. Calendar appears to point to the region of Sens. Unidentified arms, fol. 59, 59^{v}, and a partly torn book-plate: four salmon hauriant in bend, griffins' support, first pastedown.

785. **Keble College 23** Saec. XV ex.
Book of Hours. Use of Rome (200 × 146)

Fine miniatures, borders. Good and other initials. By the workshop of Jean Bourdichon. [Illus.]

Prov. 'Dawson Turner bought at Paris 1814', fol. iii. His sale, Sotheby's, 7 March 1853, lot 1881. J. D. Chambers, 1853, fol. iii. Given by Mrs. J. D. Chambers in 1893, first pastedown.

786. **Keble College 10** Saec. XV ex.
Book of Hours. Use of Troyes. (186 × 130)

Good miniatures, borders, initials.

French gold-tooled leather binding with 'Claude', panel of the Crucifixion and '1597' on the front cover; 'Huez', the Annunciation, and '1597', on the bottom cover. Resewn and rebacked, saec. XIX.

Prov. Calendar and litany suggest use in the diocese of Troyes. Claude Huez, saec. XVI. Bequeathed by Dr. Henry P. Liddon, canon of St. Paul's, London (d. 1890).

787. **Somerville College 1** Saec. XV^{2}
Book of Hours. Use of Angers (88 × 60)

Good initials. Line-fillers.

Prov. 'Ce présent livre appartient à moi Breton fille de la Vierge', saec. XVIII/XIX, fol. 78^{v}.

788. **Queen's College 305** Saec. XV^{2}
Legendary, In French (387 × 280)

Fine and good miniatures (presumably copying an earlier model). Good and other initials, cadels. Written by two hands. [Illus.]

Prov. Isaac Crommeling, 1653, fol. ii. Given by Peter Causton, London mercer, 1697, fol. iii.

P. Meyer, 'Notice du ms. 305 de Queen's College, Oxford (Legendrier français)', *Romania*, XXXIV (1905), 215–30; H. Friedman, *A Bestiary for Saint Jerome*, 1980, 311, fig. 205.

789. **Balliol College 383** 1490–1503
Ovid, Heroides. French translation of Octavien de St. Gelais. In Latin and French (336 × 230)

Good miniatures, borders, initials. Line-fillers. Miniature, fol. 3^{v}, remounted and repainted. [Illus.]

Prov. Written for Jean de Chabannes, comte de Dammartin (d. 1503), who married as his second wife *c.* 1491 Suzanne de Bourbon-Rousillon. Arms of Chabannes, Dammartin, Chatillon, fols. 3^{v}, 6^{v}, 59, 64, are impaled with arms of Bourbon, fols. 103, 161. Bequeathed by Richard Prosser, fellow, prebendary of Durham cathedral, 1839, fol. 1.

Saxl and Meier, III, 409–12.

790. **Keble College 1** *c.* 1500
Book of Hours. Use of Paris (170 × 120)

Fine miniatures, borders, initials. Miniature (coat-of-arms erased) and border added, fol. 31^v.

Prov. Calendar suggests the diocese of Paris (St. Geneviève in gold). Annals of the Dargentz family of Rennes, *c.* 1550–56, fols. 1, 150^v. Given by Miss Maria Anne Hudson, Brabyns Hall, Stockport, in 1881, fol. i^v.

791. **Queen's College 298** *c.* 1500
South France (?)
Francis Phillip (?), de homine condito (fragment) (265 × 185)

Good initial. Humanistic script. Heading, recording the gift of the MS. to Richard Fox, bp. of Winchester 1501–28, added in a different humanistic hand.

Prov. Given by Edward Phillips, author (d. 1696), to Thomas Barlow, provost, bp. of Lincoln (d. 1691), fol. i^v.

792. **Keble College 12** *c.* 1500
Rouen
Book of Hours. Use of Rouen. In Latin and French (161 × 114)

Good miniatures, borders, initials. Line-fillers. [Illus.]

Prov. 'Des Presentes (?) appartient A Marye allen femme de Martin grandin', fol. iii^v, saec. XVI ex. 'Ms francais 3', fol. ii^v. Philip Hurd sale, S. Leigh Sotheby's, 24 May 1848, lot 35. Sir William Tite sale, Sotheby, Wilkinson and Hodge's, 18 May 1874, lot 1458. Bequeathed by Dr. Henry P. Liddon, canon of St. Paul's, London (d. 1890).

793. **St. John's College 12** Saec. XVI in.
Franciscus Patricius, de regno et regis institutione. On paper (375 × 260)

Good border, initial.

Prov. Unidentified arms: azure 3 stags rampant within a border or, p. 2. Given by Henry Price, fellow (d. 1600), p. 2.

794. **Christ Church 100** Saec. XVI in.
Book of Hours. Use of Rome (117 × 80)

Fine miniatures, borders, initials. By at least two hands. [Illus.]

Prov. William Wake, archbp. of Canterbury 1715–37.

795. **Keble College 43** Saec. XVI$\frac{1}{4}$
Book of Hours. Use of Rome (165 × 120)

Fine miniatures. By two hands, the first that of the Master of the Hours of Claude de France, cf. no. 796. Borders in two styles. Good and other initials. Partly written in humanistic script. [Illus.]

Prov. Made for use in the diocese of Nantes (evidence of the calendar). Sir Thomas Brooke, Huddersfield (d. 1908). Bequeathed by his brother, Canon C. E. Brooke (d. 1911).

J. Plummer, G. Clark, *The last flowering. French painting in manuscripts 1420–1530 from American collections*, 1982, no. 128.

796. **Corpus Christi College 385–6** Saec. XVI$\frac{1}{4}$
Guyart Desmoulins, Bible historiale (Pentateuch-Esther) with gloss. In French (382 × 270)

Important and fine miniatures. By two hands, the first the Master of the Hours of Claude de France, the second of the Rouen school. Cf. no. 795. [Illus.]

Prov. Said to have been illuminated for Francis I (1515–41). Bought in Paris in 1720 by General James Oglethorpe, gentleman commoner, who gave it to the College (d. 1785).

T. Fowler, *Corpus Christi* (College Histories), 1898, 178–9.

797. **Keble College 44** 1527
Book of Hours. Use of Paris. In Latin and French (117 × 70)

Good miniatures. Borders in two styles. Good and other initials. Line-fillers. Humanistic script. [Illus.]

Binding of morocco leather gold-tooled in Duodo style with monogram 'CC' on the back cover, 'AV' on the front together with the emblem of the Holy Ghost, *c.* 1600, for Claude de l'Aubespine.

Prov. Made for Claude de Lorraine, first duc de Guise, 1527, his arms in the border, fol. 65. Claude de l'Aubespine, secretary of the Ordre du Saint Esprit 1579–1608, his monogram overpainted in border, fol. 14^v. 'Duchesse de Berry, no. 19', fol. iv. Sir Thomas Brooke, Huddersfield (d. 1908). Bequeathed by his brother, Canon C. E. Brooke (d. 1911).

L. Delisle, 'Bibliographie', *Bibliothèque de l'Ecole des Chartes*, LIII, (1892), 185, no. XIII; Exh., *Fine bindings 1500–1700 from Oxford Libraries*, Bodleian Library, 1968, no. 98.

798. **Corpus Christi College 387** *c.* 1530–40
Book of Hours. Use of Rome (278 × 180)

Fine miniatures, borders, initials. By two hands. Related

in style to Vienna, Nationalbibliothek, Cod. 1883. [Illus.]

Prov. Arms, probably de Lévis, in borders, or three chevrons sable. Given by Robert Trotman Coates, fellow, 28 May, 1833, fol. 1.

799. **Christ Church 180** 1599

Psalms. In French. On paper (187 × 130)

Good miniature, borders, historiated initials. Written in different styles and illustrated at Edinburgh in 1599 by Esther Inglis (Langlois) who also wrote Bodl. MSS. Bodley 990 and 987 (cf. Pächt, Alexander, I, nos. 864, 867). [Illus.]

Contemporary embroidered velvet binding.

Prov. Dedicated and presented to Queen Elizabeth, royal arms and emblems, fols. i^v, 81^v.

D. Laing, 'Notes relating to Mrs. Esther (Langlois or) Inglis, the celebrated calligraphist, with an enumeration of manuscript volumes written by her between the years 1586 and 1624', *Proceedings of the Society of Antiquaries of Scotland*, VI (1868), 291, no. III.

800. **Keble College 33** *c.* 1600
Nogent-le-Rotrou

Cantiques de Noel, compiled by Gabriel Hubert. In French. On paper (224 × 148)

Good miniatures, historiated and other initials. Marginal illustrations. Cadels. Apparently by the same hand as Paris, Bibl. Nat. MS. Fr. 24407, *c.* 1593–4. In different styles, presumably using different models. [Illus.]

Contemporary gold-tooled leather binding with 'Gabriel' on the front cover, 'Hubert' on the back, probably Nogent-le-Rotrou.

Prov. Gabriel Hubert, 'apoticquaire de Nogent-le-Rotrou', Eure-et-Loire, signed round the initial, fol. 94, 'GH 1598', fol. 28^v. Possibly written by him. Sale of the Mathon family, Techener, Paris, 1864, lot no. 223. Sir Thomas Brooke, Huddersfield (d. 1908). Bequeathed by his brother, Canon C. E. Brooke (d. 1911).

L. Delisle, 'Bibliographie', *Bibliothèque de l'Ecole des Chartes*, LIII (1892), 182, no. XV. E. Gautier-Desvaux, W. Koechlin-Schwartz, *Noël au Perche*, Saint-Mard-de-Réno, 1977, reproductions.

801. **Keble College 65** 1650
Paris

Office and prayers of the Virgin. In French (100 × 63)

Fine borders, initials. Written in 1650 at Paris by Nicolas Jarry (cf. Pächt, Alexander, I, no. 871).

Prov. Sir Thomas Brooke, Huddersfield (d. 1908). Bequeathed by his brother, Canon C. E. Brooke (d. 1911).

FLEMISH MANUSCRIPTS

The Josephus, Merton College MS. 317 (no. 802) is a highly important manuscript with unusual iconography and is an unexpected find in a College library. There are three good early fifteenth-century Books of Hours (nos. 805–7), of which two were made for the English market. The Bruges style associated with Willem Vrelant is represented by three manuscripts (nos. 811–3). Further examples of Flemish Hours for the English market are St. John's College MS. 82 (no. 814) and Campion Hall MS. 3 (no. 816). A fine coloured drawing of Troilus and Cressida is prefixed to a Caxton printing of Chaucer's text of 1482, St. John's College MS. 266 (no. 820). The Ordinances of the Confraternity of the Immaculate Conception, Christ Church MS. 179 (no. 823) contains a representation of Henry VII and his family. Important examples of high quality Flemish illumination of the sixteenth century are to be found in the New Testament written by Peter Meghen *c.* 1520, Corpus Christi College MSS. 13–14 (no. 825), in the Missal of the Abbey of Parc, Louvain, Wadham College MS. A.7.8 (no. 826), dated 1521, and in the Epistolary and Gospel Lectionary written also by Peter Meghen *c.* 1528–9 for Cardinal Wolsey and illuminated probably by Gerard Horenbouts, Christ Church MS. 101 (no. 827) and Magdalen College Lat. 223 (no. 828).

802. **Merton College 317** (M.2.11) Saec. XII med. Liège (?)

Josephus, Antiquitates Judaicae; de bello Judaico (410 × 300)

Fine miniatures. Fine historiated and other initials. Arabesque initials. [Illus.]

Prov. Given by Nigel de Wavere, fellow, chancellor of Oxford University 1330–2, canon of Chichester (d. 1339), fols. i^{v}, 1.

Emden, 2000; W. Cahn, 'An Illustrated Josephus from the Meuse Region in Merton College, Oxford', *Zeitschrft für Kunstgeschichte*, XXIX (1966), 295–310, figs. 1–4, 8; L. M. Ayres, 'The Tanner MS. in the Bodleian Library . . .', *J.W.C.I.* XXXII (1964), 47.

803. **Keble College 17,** fols 16^{v}–181 Saec. XIII med.

Psalter (incomplete) (131 × 90)

Good miniatures, historiated and other initials. [Illus.]

Blind-stamped pigskin over boards, metal clasps (one lost), German, saec. XVI.

Prov. Written for a member of the Franciscan order (Beatus initial with St. Francis and St. Clare, fol. 21). Adapted in Germany for Cistercian use (calendar, fols. 1–6^{v}) and bound with part of a Cistercian breviary dated 1468, colophon, fol. 203. Bequeathed by Dr. Henry P. Liddon, canon of St. Paul's, London (1890).

Frere, I, no. 543.

804. **University College 12** Saec. XIII$\frac{3}{4}$ Ghent (?)

Psalter (incomplete) (147 × 103)

Good historiated and other initials. Related in style to Bodl. MS. Liturg. 396. [Illus.]

Prov. Given by Thomas Walker, master (d.1665), fol. 2.

Hunt, *Manuscript*, 21.

805. **Jesus College 32** Saec. XV in.

Book of Hours. Use of Sarum (204 × 145)

Fine miniatures, borders (mutilated), historiated and other initials. By the same hand as Durham, Ushaw College MS. 10, from Bruges, 1408, and Bodl. MS. Laud Lat. 15. [Illus.]

Prov. Inscribed by Anthony Wood, antiquarian (d. 1695), 'Ant. a Bosco', fol. 1.

A. Clark, *ed.*, 'Wood's Life and Times', *Oxford Historical Society*, XIX (1892), 23; Pächt, Alexander, 3, no. 1312.

806. **University College 5** Saec. XV in.

Book of Hours. Use of Treguier (Dominican) (172 × 125)

Fine miniatures (partly defaced), close in style to Bodl. MS. Laud Lat. 15. Good borders, initials (cf. no. 805). [Illus.]

807. **Balliol College 384** Saec. XV$\frac{2}{4}$

Book of Hours. Use of Sarum (223 × 155)

Good miniatures, borders by two hands (inserted on single leaves). Group of the so-called Master of the Golden Scrolls (cf. no. 808). Good and other initials. [Illus.]

Prov. Obits, mainly of Lovell and Wake families, 1414–71.

808. **Corpus Christi College 18** Saec. xv$\frac{2}{4}$
Psalter (200 × 142)

Fine miniature, borders, historiated and other initials, line-fillers. Group of the so-called Master of the Golden Scrolls. Cf. no. 807. [Illus.]

Prov. Calendar, fols. 1–6^{v}, made for the diocese of Lincoln. Given by William Fullman, fellow, rector of Meysey Hampton, Glos. (d. 1688), fol. 1.

809. **Keble College 81** Saec. xv med.
Ghent (?)

Book of Hours. Use of Rome (205 × 148)

Good miniatures, borders, historiated and other initials. [Illus.]

Prov. Calendar points to Ghent. Arms on inserted leaf: azure a chevron or between in chief a representation of the Holy Spirit argent membered or bearing on the breast the letters IHS or between two five-pointed stars also or and in base a crescent argent issuant therefrom three lilies stalked or and flowered argent surrounded by two stalks of leaves with, below, a date '1557', fol. 1a. William Howley, archbp. of Canterbury (d. 1848). Rev. Christie Chetwynd Atkinson (d. 1911), fol. i.

810. **Somerville College 2** Saec. xv med.
Bruges (?)

Book of Hours. Use of Rome (108 × 76)

Good miniatures, borders, initials. [Illus.]

811. **St. John's College 68** Saec. xv$\frac{3}{4}$
Bruges (?)

Liber de regimine sanitatis. In French (290 × 190)

Fine miniature (rubbed). Fine border, initials. Cadels. Related to the style of Willem Vrelant. Cf. nos. 812, 813. [Illus.]

Prov. George Bate, M.D. from St. Edmund Hall, physician to Charles II (d. 1669).

812. **Lincoln College Lat. 149** Saec. xv$\frac{3}{4}$
Bruges (?)

Book of Hours. Use of Rome (96 × 69)

Fine miniatures, borders, initials. Related to the style of Willem Vrelant. Cf. nos. 811, 813. Italian script and parchment. [Illus.]

Fore-edge pattern, saec. XVI (?).

Prov. Translation of St. Eligius, 25 June. Given by Mr. Eric Sexton, Camden, Maine, U.S.A., in 1975.

813. **Keble College 46** Saec. xv$\frac{3}{4}$
Bruges (?)

Book of Hours. Use of Rome. In Latin and French (198 × 140)

Good miniatures, fine borders, in two styles. Good and other initials. Related to the style of Willem Vrelant. Cf. nos. 811, 812. Fol. 21 by a different hand. [Illus.]

Prov. Calendar points to the diocese of Mons. '. . . ruotin Bonus pu . . . vrbs Lugdunii civis romanus 1640', last pastedown. 'Louis Stiven 1651', first pastedown. Other owners, French, saec. XVII. Sir Thomas Brooke, Huddersfield (d. 1908). Bequeathed by his brother Canon C. E. Brooke (d. 1911).

814. **St. John's College 82** Saec. xv$\frac{3}{4}$
Book of Hours. Use of Sarum (275 × 188)

Fine miniatures, borders. Fine historiated and other initials. Line-fillers. [Illus.]

Prov. Calendar with English saints suggests London (translation of St. Erkenwald). Given by Thomas Knyvett, baron de Escrick, 1607, 'ex rogatu' of sir William Paddy, commoner, physician to James I (d. 1634), fol. 1.

815. **Christ Church 94** Saec. xv$\frac{3}{4}$
Book of Hours (incomplete). Use of Rome. In Latin and French (180 × 132)

Good miniatures, borders, initials (mutilated).

816. **Campion Hall 3** Saec. xv$\frac{3}{4}$
Book of Hours. Use of Sarum (223 × 145)

Good miniatures, borders, initials (mutilated). [Illus.]

Original stamped leather binding on boards.

Prov. Made for the English market. Given by sir Frank Brangwyn, R.A. (d. 1956).

817. **Magdalen College Lat. 252** Saec. xv$\frac{3}{4}$
Book of Hours. Use of Rome (85 × 60)

Good miniatures. Fine borders, initials. [Illus.]

Prov. Calendar points to the region of Bruges. Given by John Rigaud, fellow (d. 1888).

818. **Queen's College 349** Saec. xv ex.
Book of Hours. Use of Paris (160 × 105)

Fine miniatures, borders, historiated and other initials. Line-fillers. [Illus.]

Prov. Calendar, fols. 1–12^{v}, adapted for English use (St. Hugh, 2 feasts with octave—diocese of Lincoln (?)). Mary, daughter of Henry VII, queen of Louis XII of

France and after his death duchess of Suffolk (d. 1533). Given by the Duchess of Suffolk, her daughter, to Florence Clifford in 1550, fol. 32. Given to the College by Thomas Wilbram, possibly Queen's College B.A. in 1611, fol. iii^{v}. Leaves mutilated but returned anonymously in 1727.

819. **St. John's College 265** Saec. XV ex. Parc, Louvain

Matutinale. Premonstratensian use. On paper (195 × 135)

Good penwork border, initials.

Prov. Pressmark of Parc abbey, nr. Louvain: 'I theca XI', first pastedown, saec. XVI (?) and its arms stamped on the binding, saec. XVIII. Possibly sale of John Pecham, Slough, 5–9 May, 1831, lot 121. Given by George Hart Morley, 'commensalis', 1831, fol. 1.

820. **St. John's College 266** Saec. XV ex.

Geoffrey Chaucer, Troilus and Cressida, etc. In English. On paper (250 × 190)

Fine miniature (coloured drawing) on a single leaf preceding *Troilus and Cressida* printed by Caxton *c.* 1482, bound up with other printed texts and a MS. [Illus.]

Prov. Roger Thorney, London mercer (d. 1515), end flyleaf. Given by sir William Paddy, commoner, physician to James I (d. 1634), fol. 2.

G. Bone, 'An extant manuscript printed by W. de Worde with notes on the owner, Roger Thorney', *The Library*, XII (1932), 287-8; K. L. Scott, *The Mirroure of the Worlde* (Roxburghe Club), 1980, 32–3, fig. 19.

821. **Keble College 45** *c.* 1500

Book of Hours. Use of Rome (116 × 76)

Good miniatures, borders, initials. [Illus.]

Prov. Arms in the border, fol. 133^{v}, rubbed. Sir William Tite sale, Sotheby, Wilkinson and Hodge's, 18 May 1874, lot 1487. Sir Thomas Brooke, Huddersfield (d. 1908). Bequeathed by his brother, Canon C. E. Brooke (d. 1911).

822. **New College 323** Saec. XVI in. Flemish (?)

Book of Hours. Use of Rome (104 × 74)

Fine miniatures, borders, initials. [Illus.]

Prov. Calendar suggests Utrecht. Given in 1693 by magister Henry Howell of London.

Hunt, *Medieval library*, 339 n. 81.

823. **Christ Church 179** 1503–17

Ordinances of the Confraternity of the Immaculate Conception, London. In French (353 × 250)

Fine miniature. Fine borders, initial. [Illus.]

Prov. Portrait of Henry VII and his family with royal arms, fol. 1^{v}. Petition of the French residents in London to the king, fol. 2.

824. **University College 40** *c.* 1515

Theological pieces (225 × 155)

Fine border, fine and other initials. Humanistic script by Peter Meghen. Cf. nos. 825, 827, 828. [Illus.]

Prov. Written for Christopher Urswick, dean of Windsor (d. 1522), fol. 102.

O. Pächt, 'Holbein and Kratzer as collaborators', *Burlington Magazine*, LXXXIV (1944), 137 n. 6, pl. II D; Hunt, *Manuscript*, 26, 34; *Humanistic script* (Bodleian Picture Book no. 12), 1960, pl. 14; J. B. Trapp, 'Notes on manuscripts written by Peter Meghen', *The Book Collector*, XXIV (1975), 96, no. 23.

825. **Corpus Christi College 13–14** *c.* 1520

New Testament with the interlinear version of Erasmus (298 × 205)

Fine miniatures, borders, initials by an unidentified Flemish artist whose St. Mark is apparently based on Dürer's engraving of St. Jerome of 1514. Humanistic script by Peter Meghen (cf. nos. 824, 827, 828). [Illus.]

O. Pächt, 'Holbein and Kratzer as collaborators', *Burlington Magazine*, LXXXIV (1944), 137; J. B. Trapp, 'Notes on manuscripts written by Peter Meghen', *The Book Collector*, XXIV (1975), 88, no. 8; *id.*, 'John Colet, his manuscripts and the Ps.-Dionysius', *Classical influence on European culture A.D. 1500–1700. II. Proceedings of an international conference held at King's College, Cambridge*, ed. R. R. Bolgar, 1976, 209; H. Friedmann, *A Bestiary for Saint Jerome*, 1980, 310, fig. 200; J. B. Trapp, 'Peter Meghen, 1466/7–1540 scribe and courier', *Erasmus in English*, no. 11 (1981/2), 28–33, Appendix, 'Notes on manuscripts written by Meghen', fig. 7; H. Gibaud, *Un inédit d'Erasme; la première version du Nouveau Testament copié par Pierre Meghen 1506–1509*, 1982, 27–34, 40, 46–7, 48–9, 49–50, 54, 55–6, 57, 60, 74–5, pl. 316 (MS. 14).

826. **Wadham College A.7.8** 1521 Parc, Louvain

Missal (428 × 200)

Fine miniatures, historiated and other borders. Fine initials. Written by Francis Weert, fol. 1^{v}, who also wrote an Epistolary with similar decoration, Bodl. MS. Douce 200 *c.* 1521 (cf. Pächt, Alexander, I, 391). [Illus.]

Prov. Made for Ambrose von Angelen, abbot of Parc, nr. Louvain, fol. 1^{v}. Shelf-mark 'I (theca) XII', first pastedown, fol. 1^{v}.

Exh., *Illuminated Manuscripts*, 1908, 85, no. 172; Exh., *Flemish Art 1300–1700*, Royal Academy, London, 1953–4, no. 618.

827. **Christ Church 101** 1528

Epistolary (410 × 293)

Fine miniatures, borders, initials. By Gerard Horenbouts of Ghent. Humanistic script by Peter Meghen. Cf. nos. 824, 825, 828. [Illus.]

Prov. Cardinal Thomas Wolsey, archbp. of York 1514, bp. of Winchester 1529–30, his motto 'Dominus mihi adiutor' and his arms alone or impaling those of the see of York, incorporated in the borders.

Exh., *Flemish Art 1300–1700*, Royal Academy, London, 1953–4, no. 623; H. Paget, 'Gerard and Lucas Hornebolt in England', *Burlington Magazine*, CI (1959), 400, figs. 43, 45; J. B. Trapp, 'Notes on manuscripts written by Peter Meghen', *The Book Collector*, XXIV (1975), no. 15; Exh., *The Kings Good Servant, Sir Thomas More 1477/8–1535*, National Portrait Gallery, London, 1978, no. 43, pl.; J. B. Trapp, 'Peter Meghen 1466/7–1540 scribe and courier', *Erasmus in English*, no. 11 (1981/2), 28–33, Appendix, 'Notes on manuscripts written by Meghen', fig. 9.

828. **Magdalen College Lat. 223** *c.* 1529

Gospel Lectionary (Cardinal Wolsey's Gospels) (405 × 300)

Important miniatures, borders, initials. By Gerard Horenbouts of Ghent. Humanistic script by Peter Meghen. Cf. nos. 824, 825, 827. [Illus.]

Gold-tooled calf leather binding, arms of Henry VIII, London, saec. XVI med.

Prov. Cardinal Thomas Wolsey, archbp. of York 1514, bp. of Winchester 1529–30, his motto 'Dominus mihi adiutor', initials 'T.W.' and his arms alone or impaling those of the sees of York and of Winchester incorporated in the borders. In the King's library, February 1551. It may have been in use in Winchester Cathedral in 1556 for the administration of the oath, recorded on first pastedown. Samuel Chappington, fol. i, saec. XVI (?).

Exh., *Illuminated Manuscripts*, 1908, no. 173; Exh., *Flemish Art 1300–1700*, Royal Academy, London, 1953–54, no. 624; E. Auerbach, *Tudor Artists*, 1954, 45; H. Paget, 'Gerard and Lucas Hornebolt in England', *Burlington Magazine*, CI (1959), 400; J. B. Trapp, 'Notes on manuscripts written by Peter Meghen', *The Book Collector*, XXIV (1975), 80; *id.*, 'Peter Meghen 1466/7–1540 scribe and courier', *Erasmus in English*, no. 11 (1981/2), 28–33, Appendix, 'Notes on manuscripts written by Meghen', fig. 8.

DUTCH MANUSCRIPTS

There are rather few manuscripts in this section. The Book of Hours, Queen's College MS. 210 (no. 830) may be an example of the export of single leaves with miniatures documented as taking place from Utrecht in the fifteenth century.

829. **St. John's College 64** 1414 Netherlands (?)

Peter of Herenthals, Comm. on the Psalter. On paper (290 × 200)

Good penwork initials. Written by Gerard Teerpenniuc, 'curatus' in Lexmunde, 1414, fol. 274.

Original binding of calf leather on boards.

Prov. Ex libris of St. Albans, saec. xv, fol. 1. Given in 1617 by William Laud, president, archbp. of Canterbury 1633–41, fol. ii.

Exh., *Archbishop Laud*, no. 5; R. W. Hunt, 'The Library of the Abbey of St. Albans', *Medieval Scribes, Manuscripts and Libraries. Essays presented to N. R. Ker, eds.* M. B. Parkes, A. C. Watson, 1978, 268.

830. **Queen's College 210,** fols. 19 124^v Saec. xv$\frac{1}{4}$

Book of Hours. Use of Sarum (202 × 137)

Fine miniatures on inserted single leaves. Cf. no. 465. [Illus.]

831. **Keble College 19** Saec. xv med. Dutch or German (?)

Diurnal with prayers. Benedictine use (115 × 84)

Border, initial. [Illus.]

Prov. Written by or for a nun of a Benedictine house dedicated to St. Servatius (evidence of the calendar, etc.). Pastedowns from a document in German mentioning Recklinghausen, Westphalia, saec. xvi. Given by Rev. Greville J. Chester (d. 1892), fol. i^v.

Frere, 1, no. 545.

832. **New College 304** 1468–9 Utrecht

Hugo de Argentina, Compendium theologiae veritatis, etc. (318 × 240)

Good initials. Penwork initials. Written by Theodoricus de Huesdene, presbyter, colophon fols. 94^v, 146^v, etc.

Medieval binding of brown leather on boards, Netherlandish, saec. xv.

Prov. 'Istum librum legavit . . .' erased *ex libris*, fol. 1^v. Bequeathed by Theodoricus de Huesdene to the house of Hieronymites at Utrecht, fols. 265, 273^v.

833. **Keble College 8** Saec. xv$\frac{3}{4}$

Book of Hours. Dutch translation by Geert Grote (124 × 85)

Fine miniatures (partly defaced), borders, initials. Good penwork initials. [Illus.]

Prov. Bequeathed by Dr. Henry P. Liddon, canon of St. Paul's, London (d. 1890).

834. **Keble College 67** Saec. xv$\frac{3}{4}$

Bible (Genesis-Esdra III) (388 × 257)

Fine border, initials. Fine penwork initials by the same hand as MS. Keble 68 (no. 835). Written by the same scribe.

Black leather binding on original boards, trace of blind tooling, original sewing. Metal clasps now lost.

Prov. 'Ist(e liber) . . . In (Delft)' (?), blotted out inscription, fol. 213. Bequeathed by Dr. Henry P. Liddon, canon of St. Paul's, London (d. 1890).

835. **Keble College 68** Saec. xv$\frac{3}{4}$

Bible (Proverbs—II Macc.) (388 × 257)

Fine penwork initials. By the same hand as MS. Keble 67 (no. 834). Written by the same scribe.

Prov. Cf. no. 834.

836. **St. Hilda's College s.n.** *c.* 1475–80 Utrecht

Book of hours. Use of Utrecht. In Dutch (193 × 123)

Fine historiated border, historiated initials. Good penwork initials. [Illus.]

Prov. Calendar points to Utrecht. Various unidentified coats of arms. Given by Miss Helen Smith (d. 1944).

837. **Keble College 77** Saec. XV ex. Utrecht

Book of Hours. In Dutch (163 × 113)

Fine miniatures, borders, on single leaves. Fine borders, initials by a different hand. Fine penwork borders, initials. [Illus.]

Original boards and sewing with stained parchment cover.

Prov. Calendar of Utrecht. Book-plates of Madame van Huerne de Puyenbeke née de Schietere de Lophem, saec. XIX. William Morris (d. 1896). Rev. Christie Chetwynd Atkinson (d. 1911).

838. **Keble College 79** Saec. XV ex.

Book of Hours. Use of Rome. In Latin and Dutch (173 × 119)

Good miniatures, borders. Good and other initials. Penwork initials.

Prov. Rev. Christie Chetwynd Atkinson (d. 1911), fol. i.

839. **Trinity College 89** Saec. XV² Dutch or English

Miscellaneous prayers, offices, meditations, etc. (400 × 300)

Good historiated and other initials in unusual style, by two hands. Cadels. [Illus.]

840. **Keble College 28** 1537

Rule of St. Augustine, etc. In Dutch (232 × 178)

Good border, historiated initial. Written in 1537 by Fr. Herman Van Doesborch, colophon fol. 1ᵛ. [Illus.]

Stamped leather binding, original boards and sewing, metal bosses, saec. XVI. Rebacked.

Prov. Bethlehem Augustinian convent at Hoorn (professions and obits of nuns in the calendar, fols. 1–19ᵛ, statutes of the convent, fols. 36–62). I. Enschedé sale, F. Muller & Nijhoff's, Haarlem, 1867, lot 732. William Bragge sale, Sotheby, Wilkinson & Hodge's, 1876, lot 22. Sir Thomas Brooke, Huddersfield (d. 1908). Bequeathed by his brother Canon C. E. Brooke (d. 1911).

GERMAN, HUNGARIAN AND BOHEMIAN MANUSCRIPTS

The Regensburg Lectionary, Keble College MS. 49 (no. 841) is well known and is the most important manuscript in this section. Another fine liturgical manuscript from Keble College, is the Cistercian Missal, MS. 54 (no. 844). The *Secreta Secretorum* made for Louis I, King of Hungary, Hertford College MS. 2 (no. 842) and the Hours of Wenceslas of Bohemia, Pembroke College MS. 20 (no. 846) are two unexpected manuscripts to find in College libraries, and how they came to find their way to England is a mystery. A large number of manuscripts given to Balliol College by William Gray were decorated for him in Cologne (nos. 848, etc.). The Book of Hours, Keble College MS. 35 (no. 872) and a fragment probably of an Hours, Campion Hall MS. 1 (no. 873) are both of fine quality, probably coming from south Germany, but needing further identification.

841. **Keble College 49** — *c.* 1267–76, Regensburg

Lectionary (442 × 305)

Fine miniatures, historiated and other initials. Good penwork initials. Written by two hands. Textile covers for the initials. [Illus.]

Binding of white leather over boards, chemise, brass centre and corner pieces. Centre inscribed 'O Maria hilf Maria in demm himmel'. German, saec. XVI.

Prov. Made for the convent of Dominican nuns of the Holy Cross at Regensburg, perhaps at the expense of Gebehard von Hierzpech and his wife (married in 1258), represented adoring Christ crucified, fol. 7. Datable on liturgical evidence: revision of liturgy introduced by the bull of Clement IV in 1267, and the feast of Martha established 1276, added by a later hand on fol. 307v. Sir Thomas Brooke, Huddersfield, (d. 1908). Bequeathed by his brother, Canon C. E. Brooke (d. 1911).

H. Swarzenski, *Die deutsche Buchmalerei des XIII. Jahrhunderts*, 1936, no. 27, Taf. 61–68.

842. **Hertford College 2** — 1370–82, Hungary

Ps.-Aristotle, Secretum secretorum (240 × 173)

Fine historiated initial by the Master of the Hungarian Illuminated Chronicle, probably Nicolaus Meggyesi, son of Hertful the court painter. Faint marginal sketches, saec. XIV/XV (?), fols. 7, 48, 50, etc. [Illus.]

Prov. Made for Louis I, king of Hungary 1342, and of Poland 1370 (d. 1382), shown in the initial with his arms and the arms of Poland below, fol. 1. Given by Edmund Hall, fellow of Pembroke College, to Magdalen Hall (Hertford College) in 1658, fol. 1.

Exh., *Hungarian Art Treasures*, Victoria and Albert Museum, London, 1967, no. 137; D. Dercsényi, *The Hungarian Illuminated Chronicle*, 1969, 44, 57 n. 83, pl. 10; C. Csapodi, *The Corvinian Library*, 1973, 140, no. 67.

843. **Keble College 69** — Saec. XIV in., West Germany or Flanders

New Testament (441 × 317)

Fine historiated and other initials. Penwork initials. [Illus.]

Prov. Probably made for a member of a Franciscan house who kneels before St. Matthew, fol. 1. Bequeathed by Dr. Henry P. Liddon, canon of St. Paul's, London (d. 1890).

844. **Keble College 54** — Saec. XIV in., Cologne (?)

Missal. Cistercian use (351 × 267)

Fine miniature, border. Fine and other penwork initials by a different hand. [Illus.]

Prov. Written for a Cistercian house probably in Germany. Contents (collect, secret and postcommunion for St. Cornelius) suggest diocese of Cologne. William Bragge sale, Sotheby, Wilkinson and Hodge's, 7 June 1876, lot 115. Sir Thomas Brooke, Huddersfield (d. 1908). Bequeathed by his brother, Canon C. E. Brooke (d. 1911).

845. **Balliol College 274** 1409
German

John of Wales. Communiloquium, etc. (incomplete). (330 × 210)

Fine initials (mutilated). Written by Paulus de Mertzenich, colophon fols. 130, 165, 235^{v}, 254^{v}.

Prov. Probably acquired by William Gray, bp. of Ely 1454–78, in Cologne *c.* 1442–4.

L. Thorndike, 'All the world's a chess board', *Speculum*, VI (1931), 461–5.

846. **Pembroke College 20** *c.* 1410–19
Prague

Book of Hours (163 × 120)

Fine borders, historiated initials. Related in style to the Joshua Master of the Bible of Konrad von Vechta. Initials, fols. 33^{v}, 47^{v}, in a different style close to the Master of the Gerona Martyrology (cf. Alexander, 31). [Illus.]

Blind-stamped leather panel binding rebacked, Netherlandish, saec. XV/XVI.

Prov. Made for Wenceslaus IV, king of Bohemia (d. 1419), his initial 'W' in the prayers, fols. 14^{v}, 16^{v}, 24^{v}, and probably his portrait with Sophia of Bavaria, fol. 46. Franciscus Muijlwijch, fol. 1, saec. XVI.

J. J. G. Alexander, 'A Book of Hours made for King Wenceslaus IV of Bohemia', *Studies in Late Medieval and Renaissance Painting in Honor of Millard Meiss, eds.* I. Lavin, J. Plummer, 1978, 28–31, figs. 1–7.

847. **Balliol College 31** 1432
Cologne (?)

Augustinus de Ancona, Comm. on Canonical Epistles, etc., with gloss (295 × 210)

Fine historiated and other initials (mutilated). Penwork initials. Written by a probably German scribe, colophon fol. 189^{v}. [Illus.]

Prov. William Gray, bp. of Ely 1454–78, fol. i^{v}.

848. **Balliol College 78A** 1442
Cologne

Thomas Waleys, Nicholas Trivet, Comm. on Augustine, de civitate Dei (355 × 255)

Fine borders (mutilated), initials. Fols. 1–3 by the English illuminator of MS. Balliol 276 (no. 442), the remainder by the German artist of MS. 238A (no. 863). Written in Cologne by Richard Bole, colophon and his motto 'Abide and Have', fol. 118^{v}, cf. no. 517. For other MSS. see Index.

Prov. William Gray, bp. of Ely 1454–78, fol. i^{v}.

Mynors, *Catalogue*, xlviii.

849. **Balliol College 265** 1442–3
Cologne

Alan of Lille, Anticlaudianus (210 × 138)

Good border (mutilated).

Prov. Nicholas Saxton, fellow, rector of Danbury, Essex, 1461–71, fol. 2^{v}.

Emden, 1648.

850. **Balliol College 276,** fols. 57–154^{v} 1442–44
Cologne

Terence, Comedies, etc. (340 × 255)

Initials in Rhenish style added to a MS. written for William Gray in Oxford, cf. no. 442.

Prov. William Gray, bp. of Ely 1454–78.

Mynors, *Catalogue*, xxvii; Exh., *Duke Humfrey*, 25–6.

851. **Balliol College 279** *c.* 1442–4
Cologne

Dictionarium (damaged by damp) (380 × 240)

Good border (mutilated), initial.

Prov. Possibly written for William Gray in Oxford, illuminated in Cologne *c.* 1442–4.

Mynors, *Catalogue*, xxvii.

852. **Balliol College 180** *c.* 1442–4
Cologne

Bertrandus de Turra, Sermons (370 × 277)

Fine border (mutilated). Written by a German or Dutch scribe probably *c.* 1442–44.

Prov. William Gray, bp. of Ely 1454–78, fol. 1^{v}. List of contents in the humanistic hand of Richard Bole, fol. 1^{v}.

Exh., *Duke Humfrey*, 26.

853. **Balliol College 79** *c.* 1442–4
Cologne

Lactantius, Tertullian (370 × 246)

Good border (mutilated), initials. Penwork borders, initials. [Illus.]

Prov. William Gray, bp. of Ely 1454–78, fol. 2^{v}.

854. **Balliol College 129** fols. 235 col. 2 line 3–252v
c. 1442–4
Cologne (?)

Seneca, Various works (348 × 240)

Border, penwork initial in Rhenish style. Additions to a MS. of saec. XIII (?). Written by a German scribe probably at Cologne. Cf. no. 257.

Prov. Ex libris of Balliol College, saec. XIV ex., fol. 4v of the original MS. and note by Thomas Gascoigne, chancellor of Oxford University (d. 1458), fol. 8. William Gray, bp. of Ely 1454–78, fol. 4v. Annotations by Richard Bole, fellow, archdeacon of Ely (d. 1477).

855. **Balliol College 55** *c.* 1442–4 (?)
Rhineland (Cologne (?))

Gerard de Senis, Comm. on the Sentences (350 × 273)

Border (mutilated).

Prov. William Gray, bp. of Ely 1454–78, fol. ivv.

856. **Balliol College 125** 1442–4
Cologne

Eutropius, de gestis Romanorum (324 × 235)

Good border, initials (mutilated). By the illuminator of MS. Balliol 238D (no. 866) also written for William Gray in Cologne by the same scribe, probably Richard Bole, who wrote MS. Balliol 78A for Gray (no. 848).

Prov. Parchment label from the old cover with arms of William Gray, bp. of Ely 1454–78, pasted on fol. 3v.

Mynors, *Catalogue*, xxix; *id.*, *Werken*, 99; de la Mare, *A fragment*, 285 n. 3.

857. **Balliol College 130** *c.* 1442–4
Cologne (?)

Seneca, Various works (358 × 225)

Good penwork borders, initials. Two manuscripts bound together, the second, fols. 94–141v, signed by Henricus Horn of Düsseldorf. Humanistic script.

Prov. William Gray, bp. of Ely 1454–78, fol. 2v. Annotations and list of contents by Richard Bole, archdeacon of Ely (d. 1477).

P. T. Eden, 'William Gray, bishop of Ely, and three Oxford MSS. of Seneca', *Classica et Medievalia*, XXI (1960), 29–35; Exh., *Duke Humfrey*, no. 41, pl. VIII (a); Pächt, Alexander, 3, no. 1085.

858. **Balliol College 58** *c.* 1442–4 (?)
Cologne (?)

Tabulae (Henricus Koflin), etc. (267 × 185)

Good penwork borders, initials.

Prov. 'Liber Gulielmi Gray et Amicorum', fol. i. William Gray, bp. of Ely 1454–78, fol. iv.

859. **Balliol College 224B** 1444
Cologne

Bartoldus de Mosburch, Comm. on Proclus. Elementatio theologica (370 × 280)

Good border (mutilated), initials. Written for William Gray by a German scribe in Cologne, 1444, fol. 325v.

Prov. William Gray, bp. of Ely 1454–78, fol. 1v.

860. **Balliol College 181** 1444
Cologne

Johannes de Hesdinio, Comm. on St. Mark, etc. On paper (410 × 290)

Fine initials. Good penwork border, initials. Written probably by two Dutch or German scribes at Cologne, 1444, colophon fol. 169v.

Prov. William Gray, bp. of Ely 1454–78, fol. 1v.

861. **Balliol College 66** *c.* 1444
Cologne

Franciscus de Mayronis, Sermones hiemales (365 × 270)

Border, initial (unfinished and mutilated). Good penwork initials. Written for William Gray in Cologne by Theodericus Werken, fol. 254v. Cf. no. 447. For other MSS. see Index.

Prov. William Gray, bp. of Ely 1454–78, fol. 1v.

Mynors, *Werken*, 98, 104; *id.*, *Catalogue*, xxix; de la Mare, *A fragment*, 285 n. 4.

862. **Balliol College 67A** *c.* 1444
Cologne

Franciscus de Mayronis, Sermones aestivales (370 × 270)

Good border, initial (unfinished). Good penwork initials. Marginal drawings by the scribe, fols. 185v, 197v. Written for William Gray in Cologne by Theodoricus Werken, fols. 206, 252v. Cf. no. 447. For other MSS. see Index.

Prov. William Gray, bp. of Ely 1454–78, fol. iv.

Mynors, *Werken*, 98, 104; *id.*, *Catalogue*, xxix; de la Mare, *A fragment*, 285 nn. 4, 5.

863. **Balliol College 238A** *c.* 1444–5
Cologne

Domenico Bandini, Fons memorabilium universi, I (350 × 244)

Fine borders, initials. Marginal sketch, fol. 47. Cadels. Written for William Gray in Cologne by Theodericus Werken. Cf. no. 447. For other MSS. see Index.

Prov. William Gray, bp. of Ely 1454–78, his arms in the border, fol. 1.

Mynors, *Werken*, 99, 104; Exh., *Duke Humfrey*, pl. XXIV (f); de la Mare, *A fragment*, 285 n. 7.

864. **Balliol College 238B** 1445–8
Cologne

Domenico Bandini, Fons memorabilium universi, II–III (350 × 244)

Fine initials by two hands, the second, fols. 79v–192, perhaps of an English artist imitating the style of those earlier in the volume. Marginal drawings by scribe, fols. 3v, 27, 74, etc. Written for William Gray in Cologne by Theodericus Werken up to fol. 79v, completed by him in Rome in 1448, colophon fol. 192v. Cf. no. 447. For other MSS. see Index.

Prov. William Gray, bp. of Ely 1454–78, fol. 1v.

Mynors, *Werken*, 99, 104; de la Mare, *A fragment*, 285 nn. 3, 6.

865. **Balliol College 238C** *c.* 1445
Cologne

Domenico Bandini, Fons memorabilium universi, IVA (350 × 244)

Fine borders, initials, in the style of no. 863. Written for William Gray in Cologne by Laurentius Dyamas, the same scribe as MS. Balliol 238E, cf. nos. 867, 1004.

Prov. William Gray, bp. of Ely 1454–78, fol. 2v.

Mynors, *Catalogue*, xxvii; *id.*, *Werken*, 99.

866. **Balliol College 238D** *c.* 1445
Cologne

Domenico Bandini, Fons memorabilium universi, IVB (350 × 244)

Fine border, initials. By the illuminator of MS. Balliol 125 (no. 856). Written by the same scribe showing humanistic influence, probably Richard Bole, cf. no. 517. For other MSS. see Index.

Prov. William Gray, bp. of Ely 1454–78, his arms in the border, fol. 1.

Mynors, *Werken*, 99, 104; de la Mare *A fragment*, 285 n. 3.

867. **Balliol College 238E** *c.* 1445
Cologne

Domenico Bandini, Fons memorabilium universi, V (357 × 246)

Fine border (mutilated), initials. Cadels. Fine marginal drawings by the scribe, Laurentius Dyamas, who signed on fol. 162, for which see no. 1004. Cf. no. 865.

Prov. William Gray, bp. of Ely 1454–78, fol. 2v.

Mynors, *Werken*, 99.

868. **Hertford College 41,** fols. 45–47v
Saec. xv med.
German

Unidentified, Quaestiones in Aristotle, de anima, etc. (fragment) (*c.* 287 × 205)

Initials.

869. **Corpus Christi College 101** 1472
Bohemian

Ptolemy, Geographia, Astronomica (300 × 225)

Good borders, initials. 'Per me Jeronimum de Sprottana A.D. 1472', fol. 250v.

Prov. Possibly Dr. John Dee (d. 1608).

James, *Dr. John Dee*, 36.

870. **Corpus Christi College 219** *c.* 1484
Prague (?)

Statius, Thebais with gloss, etc. (211 × 150)

Good initials. Cf. Bodl. MS. Laud Misc. 516 (Pächt, Alexander, I, no. 161). Written in the College of king Wenceslas of Bohemia 1476–84, 'quem scripsit et terminavit Andreas Strangel de Nahradow', colophon fols. 309, 310.

871. **Keble College 18** Saec. xv²
German

Prayers and meditations. In Latin and Low German (114 × 85)

Historiated and other initials. Marginal drawings (coloured), fols. 7, 55v, 100.

Prov. Written by or for a Benedictine nun. Vernacular prayers in Low German and devotion to St. Maurice might suggest the diocese of Magdeburg. 'V. Wendt Bishof zu Pasinople (?) Vicarius Apostolicus Domprobst zu Hildesheim Conmandeur (. . .) Preussischen Grossen rothen Adeler Ordens', fol. ii, saec. XVIII. Bequeathed by Dr. Henry P. Liddon, canon of St. Paul's, London (d. 1890).

Frere, I, no. 544.

872. **Keble College 35** Saec. xv/xvi
South Germany (?)

Prayers and meditations. In German (247 × 173)

Important and fine miniatures, borders. Fine historiated (one) and other initials. By three hands. [Illus.]

Prov. Prayers in Franconian dialect. Made for a lady shown in the initial on fol. 2v with her arms: argent a cinquefoil gules pierced azure. Same arms on fol. 2 carry a crest: a demi bishop mitred. Arms on fol. 1v (inserted leaf): quarterly, 1 and 4 per fess or a demi eagle displayed sable and azure three cinquefoils argent, 2 and 3 barry of four agent and gules, with a crest: out of a coronet or a demi eagle proper crowned or and issuing from the crown two spears with banners barry of six argent and gules. Family chronicle compiled by Maria Sidonia, countess of Falckenstein in the Palatinate, added on fols. 54–7^{v}. William Bragge sale, Sotheby, Wilkinson and Hodge's, 7 June 1876, lot 52. Sir Thomas Brooke, Huddersfield (d. 1908). Bequeathed by his brother, Canon C. E. Brooke (d. 1911).

873. **Campion Hall 1** Saec. XVI in.
South Germany

Prayers (fragment of a Book of Hours (?)) (146 × 113)

Fine miniatures, one (rubbed) related to a composition of Albrecht Altdorfer. Borders, initial. [Illus.]

Prov. Given by B. Bisgood (d. 1968).

ITALIAN MANUSCRIPTS

It is perhaps surprising to find seven Italian manuscripts of the eleventh–twelfth centuries in the College libraries. The Haimo from S. Lorenzo fuori le Mura, Rome, Balliol College MS. 183 (no. 874) was no doubt acquired in Rome by William Gray, and the Terence, Brasenose College MS. 18 (no. 876) belonged to Henry Wotton, ambassador in Venice. The important copy of Joachim of Fiore's *Figurae*, Corpus Christi College MS. 255A (no. 882) is well known. Bolognese fourteenth-century illumination is well represented by All Souls College MS. 52 (no. 904), Justinian, Digest, and by New College MS. 184 (no. 907), Gregory IX, Decretals. Corpus Christi College MS. 410 (no. 908) is the important early Ps.-Bonaventura, *Meditationes*, which has a large number of miniatures very few of which have been been reproduced.

In the fifteenth-century sections Florentine white-vine illumination is well represented, many of the manuscripts being by scribes who signed their names or who have been identified by Dr. de la Mare. A few are attributable to known illuminators, for example, University College MS. 54 (no. 956), John Chrysostom, attributable to Francesco d'Antonio del Cherico.

The most important of the Roman and south Italian manuscripts is the Monte Cassino Breviary dated 1404, Keble College MS. 30 (no. 964). An unpublished Psalter, St. John's College MS. 131 (no. 967) can be attributed to the Neapolitan illuminator, Cola Rapicano. In the north-east Italian section Keble College MS. 31 (no. 972), a Breviary, is clearly by Cristoforo Cortese and two earlier manuscripts may be connected with him also, New College MS. 155 (no. 909) and Keble College MS. 59 (no. 910). University College MS. 94 (no. 988) is a good example of Ferrarese illumination. In the north-west Italian section, Lady Margaret Hall Borough MS. 1 (no. 997) is a typical example of the style of the so-called 'Ippolita Master', whilst the Landriani Hours, Keble College MS. 63 (no. 999) by the Master of the Arcimboldi Missal is an important known manuscript.

I. ITALIAN MANUSCRIPTS TO 1400

874. **Balliol College 183** Saec. XI ex.
San Lorenzo fuori le Mura, Rome

Haimo, Comm. on St. Paul's Epistles (398 × 267)

Good and other initials. [Illus.]

Prov. Written at and for the monastery of San Lorenzo ('Lorenzo' in capital letters, fols. 30, 107). Pledges and notes referring to San Lorenzo, 1136 and later, fol. 207^{v}. William Gray, bp. of Ely 1454–78, fol. 1^{v}.

875. **Balliol College 367** Saec. XI^{2}

Antidotarium (incomplete) (148 × 100)

Initial. Good marginal drawings, fols. 8, 12^{v}, 16^{v}, saec. XI/XII (mutilated).

Prov. Sir John Conroy (d. 1900) from W. Bliss, Rome, 1895, fol. i^{v}.

876. **Brasenose College 18** Saec. XI–XII

Terence, Comedies (153 × 100)

Fine initial. [Illus.]

Prov. Acquired in Venice in 1632 from the heirs of Cardinal Bembo (d. 1547) by sir Henry Wotton, ambassador, who gave it to William Hacwell, fol. iii^{v}. Given to the College by Robert Shippen, principal, 1738, fol. i.

877. **Magdalen College Lat. 50** Saec. XII in.
Central Italy (?)

Solinus, de situ orbis terrarum, etc. (220 × 145)

Good initial.

878. **Magdalen College Lat. 113** Saec. XII$\frac{3}{4}$
Umbria (?)

Peter Lombard, Comm. on the Psalter (345 × 215)

Fine historiated initials. Good arabesque initials. [Illus.]

Prov. Annotated by Andrew Holes, archdeacon of Wells (d. 1470).

879. **New College 60,** fols. 241–43ᵛ Saec. XII ex. Central Italy

Comm. on Gratian, Decretum (fragment) (390 × 245)

Fine and other initials. A fragment bound up as fly-leaves in a MS. of Peter Lombard, Comm. on the Psalter, *c.* 1250, Italy, with English penwork initials added, saec. XIII². [Illus.]

Prov. William Warham, fellow, archbp. of Canterbury 1503–32.

Emden, 1990.

880. **Merton College 263** (C.2.14) *c.* 1200 Italy or South France

Isaac Judaeus, de dietis universalibus, etc. (172 × 135)

Initials (mutilated). Penwork initials.

Prov. 'Iste liber . . . Willelmus de . . . Consobrino suo . . . eiusdem ordinis', fol. 1, saec. XIII² (?).

881. **Keble College 55** Saec. XII/XIII Morimondo

Missal. Cistercian use (276 × 192)

Initials.

Prov. List of relics in the abbey of Morimondo, nr. Milan, in 1292, added on fols. 158ᵛ–9. Liturgical MSS. sale, Sotheby, Wilkinson and Hodge's, 10 Nov. 1880, lot 1331. Sir Thomas Brooke, Huddersfield (d. 1908). Bequeathed by his brother, Canon C. E. Brooke (d. 1911).

882. **Corpus Christi College 255A,** fols. 4–17 *c.* 1200–2 South Italy

Joachim of Fiore, Figurae (340 × 230)

Good diagrams, tables. Fine genealogical trees. [Illus.]

Prov. Bequeathed by Brian Twyne, fellow (d. 1644).

F. Saxl. 'A spiritual encyclopedia of the later Middle Ages', *J.W.C.I.* v (1942), 108 ns. 1, 5, pl. 27a; L. Tondelli, *Il Libro delle Figure*, 1953, I, *passim*, II, 5–8, 15, 18 ff., pls. XIa, XX–XXIII; K. Hoffmann, *The Year 1200*, New York, Metropolitan Museum, 1970, 291, no. 285; P. Brieger, M. Meiss, C. Singleton, *Illuminated Manuscripts of the Divine Comedy*, I, 1970, 112, fig. 129; M. Reeves, B. Hirsch-Reich, *eds.*, *The 'Figurae' of Joachim da Fiore*, 1972, v–viii, 76–9 and *passim*, figs. 10–18, 20–1, 23, 25–9, 32.

883. **Keble College 21** Saec. XIII in. Milan (?)

Gospel of St. John, with gloss (231 × 149)

Good initials. [Illus.]

Prov. Erased *ex libris* 'Iste liber est conventus fratrum ordinis predicatorum sancti (Eustorgii) Mediolanensis vol. VII° Johannis', Shelf-mark '7BB' crossed through, fol. ii. Bequeathed by Dr. Henry P. Liddon, canon of St. Paul's, London (d. 1890)

884. **Oriel College 22** Saec. XIII$\frac{2}{4}$ Bologna (?)

Justinian, Institutiones with gloss of Accursius (412 × 239)

Fine historiated initial. Fine penwork initials. Marginal drawing, fol. 2. [Illus.]

885. **New College 205** Saec. XIII$\frac{2}{4}$ Bologna (?)

Gregory IX, Decretals with gloss of Bartholomeus de Segusio (400 × 275)

Good initials, fols. 30–241. Pen-drawings on fols. 1–29, written in a different Italian hand. Good initial in French style added *c.* 1300, fol. 2.

Prov. Given by John Frende, fellow (d. 1492), to William Warham, fellow, archbp. of Canterbury 1503–32. William Bodicote, fellow (d. 1513), fol. 241.

Emden, 209, 727, 1207.

886. **All Souls College 68**, fols. 113–183ᵛ Saec. XIII¹ Padua (?)

Comm. on Galen, de sanitatis ingenio, etc. (290 × 185)

Good initials. Bound with medical miscellanea written in France, saec. XIII ex.

Medieval boards under later leather cover.

Prov. Pledged in the Lincoln chest, Oxford University, 1326, 1327, fol. 295ᵛ.

Exh., *Oxford College Libraries in 1556*, Bodleian Library, Oxford, 1956, no. 45.

887. **Corpus Christi College 275,** fols. 1–112 Saec. XIII¹ North Italy (?)

Isaac Judaeus, de dietis universalibus, etc., with gloss (218 × 155)

Good initials. Penwork initials.

Prov. Cf. no. 253.

888. **New College 176,** fols. 236–237ᵛ (fragment) Saec. XIII¹ Bologna (?)

Comm. on Gratian, Decretum (390 × 245)

Initials (rubbed). End-leaves added to a contemporary Italian MS. containing penwork initials (unfinished).

Prov. W. North, fellow, rector of Droxford, Hants. (d. 1461), fol. i^{v}. William Warham, fellow, archbp. of Canterbury 1503–32, fol. 236.

889. **All Souls College 55** Saec. XIII med. Padua (?)

Azo, Summa. Roffredus de Benevento, de ordine iudiciorum (370 × 228)

Fine historiated and other initials. Fine marginal drawings. Good penwork initials by two hands. [Illus.]

Prov. Pledged in the Queen's chest, Oxford University, in 13.., fol. 1. Bequeathed by John Lyndfeld, dean of the Arches (d. 1440), fol. 1^{v}.

890. **Radcliffe Trust d. 60** Saec. XIII med. North-east Italy

Avicenna, Canon (incomplete). Latin translation by Gerard of Cremona (260 × 180)

Fine initials.

Prov. The MS., formerly Radcliffe Science Libr. RR.x.169, was recently transformed to Bodleian Library.

891. **Magdalen College Lat. 216** Saec. XIII med. Bologna

Bartholomew of Brescia, Concordantia discordantium canonum with gloss (365 × 295)

Good penwork initials. Italian script and parchment. Fine initials (mutilated) added in England, saec. XIII$\frac{3}{4}$.

Medieval binding of white leather on boards.

892. **Lincoln College Lat. 95** Saec. XIII$\frac{3}{4}$ Naples

Thomas Aquinas, Comm. on the Sentences. Bk. I (280 × 200)

Good historiated initial. Penwork initials. [Illus.]

Prov. Richard Chester, fellow of Merton College (d. before 1464), fols. 2^{v}, 5. Given by William Chamberlayne, first rector of Lincoln College (d. 1434), fol. 2^{v}.

Weiss, *Catalogues*, 351 (124) n. 140; A. Dondaine, 'Autour des secretaires de St. Thomas', *Miscellanea Medievalia, Band 2. Die Metaphysik im Mittelalter* (Veröffentlichungen des Thomas-Instituts an der Universität Köln, vorausgegeben von Paul Wilpert), 1963, 764.

893. **Balliol College 281** Saec. XIII ex.

Aegidius Beneventanus, Ortus copiosus, etc. (276 × 203)

Historiated initial (rubbed). Penwork initials.

Prov. Guard-leaves, fols. 193–4, are rejects from a collection of statutes from Todi (?), saec. XIII^{2}. William Gray, bp. of Ely 1454–78, fol. 1.

894. **Keble College 2** Saec. XIII ex. Bologna (?)

Raymundus de Penafort, Summa de penitentia (276 × 203)

Good borders, historiated (one) and other initials. Penwork initials.

Prov. Given by Rev. Greville J. Chester in 1882, fol. i.

895. **Balliol College 2**, fols. 1–120 Saec. XIII ex. Bologna (?)

Bible (267 × 192)

Fine historiated and other initials (cf. no. 700). [Illus.]

Prov. Richard Coling, armiger, 1677, fol. 1.

896. **Corpus Christi College 73,** fols. 1–225^{v} Saec. XIII^{2} Bologna (?)

Justinian, Institutiones, etc., with gloss of Accursius (415 × 250)

Good historiated and other initials. [Illus.]

Prov. Deposited by magister John Swinfield as caution for Great Stockwell Hall, 31 Jan 1297. Given by the founder, Richard Foxe, bp. of Winchester 1501–28, fol. 1.

Emden, 717; A. B. Emden, 'Additions and corrections to *A Biographical Register*...', *B.L.R.* VII (1964), 160.

897. **Merton College 264** (C.2.15) *c.* 1300 Italian, North (?)

Guillaume Peyraut, de virtutibus (204 × 151)

Good and other initials. Good penwork initials.

Prov. Given by Richard FitzJames, warden 1483–1507, bp. of London (d. 1522), who bought it in 1482, fols. iv^{v}, 264^{v}.

898. **New College 157** *c.* 1300

Anonymous, de moralitatibus rerum (320 × 230)

Fine border, historiated and other initials. Imitating French style. Penwork initials. [Illus.]

899. **New College 124** *c.* 1300 Naples

Thomas Aquinas, Summa theologiae, tertia pars (382 × 234)

Fine border, historiated initial. Good penwork initials. *Pecia* marks. Copied from a University of Naples exemplar. [Illus.]

Prov. Pledged in the Vaughan chest, Oxford University, for loan by John Echingham, 14 October 1348. Sold by Thomas Trillek, bp. of Rochester (d. 1372), to William Reed, bp. of Chichester 1368–85, who gave it to the College.

R. T. Gunther, 'Early Science in Oxford. II', *Oxford Historical Society*, LXXIV (1923), 381; J. Destrez, *La Pecia dans les manuscrits universitaires de XIIIe et de XIVe siècle*, Paris, 1935, 101, pls. 35, 36; G. Pollard, 'The University and the book trade in medieval Oxford', *Miscellanea Medievalia*, III (1964), 343.

900. **Merton College 300** (A.3.1), fols. 1–156 *c.* 1300 Italian, Central (?)

Seneca, Misc. works (423 × 270)

Fine borders, historiated (one) and other initials. Written by Gerard Johannes of Remiremont, Toul diocese, fol. 156. Cf. no. 772. [Illus.]

Prov. John de Beauvais, bp. of Angers (d. 1479), fol. i^{v}. Came to Merton College in 1494 'ex proquisitione' of Richard FitzJames, warden 1483–1507, bp. of London (d. 1522), fol. i^{v}.

901. **Lincoln College Lat. 76,** fols. 1–42 Saec. XIV in. North Italy

Valerius Maximus, Memorabilia (248 × 180)

Border (mutilated). Penwork initials in various styles.

Prov. Given by Robert Flemmyng, dean of Lincoln (d. 1483), fol. 1.

902. **Balliol College 102** Saec. XIV in. Bologna

Albertus Magnus, Comm. on Aristotle, Physics (310 × 210)

Initials (mutilated). Diagrams.

Prov. William Gray, bp. of Ely 1454–78, fol. 2^{v}.

903. **Merton College 229** (N.3.10) Saec. XIV$\frac{1}{4}$ Bologna

Avenzoar, Practica medicinae, etc. (384 × 235)

Fine borders, historiated and other initials (mutilated). Good penwork initials. [Illus.]

Prov. 'Laudi Petri de colle valles e . . . . valles else', fol. i. Given by Richard FitzJames, warden 1483–1507, bp. of London (d. 1522), fol. 1^{v}.

Emden, 692.

904. **All Souls College 52** Saec. XIV$\frac{1}{4}$ Bologna

Justinian, Digestum novum with gloss of Accursius (450 × 280)

Fine historiated and other initials (mutilated). Fine penwork initials. 'Ego Magister Mutinensis scriptor. Diana salutem etc.', fol. 299. [Illus.]

White leather on boards, saec. XV/XVI (?).

Prov. 'Dominus Bonifatius de Galluciis doctor legum in iure canonico', fol. 299, saec. XIV.

905. **Magdalen College Lat. 142** Saec. XIV^{1} Bologna (?)

Johannes Eben Mesue, Medical treatises (395 × 245)

Good border (mutilated), initials. Good penwork initials.

Prov. Jacobus de Dominis, physician, saec. XV, fol. 113^{v}.

906. **Merton College 228** (N.1.2) Saec. XIV^{1} Padua (?)

Razes, Divisiones; Almanzor, etc. (285 × 180)

Good initials.

907. **New College 184** Saec. XIV med. Bologna

Gregory IX, Decretals with gloss (465 × 275)

Fine historiated and other initials. Good penwork initials. [Illus.]

Prov. Given by William Deighton, canon of St. Paul's, London, *c.* 1371–91, fol. 319^{v}.

908. **Corpus Christi College 410** Saec. XIV med. Central Italy

Ps.-Bonaventura, Meditations on the life of Christ (217 × 150)

Fine miniatures in two styles. [Illus.]

Prov. Possibly copied for a lady, her likeness and arms (obliterated) on fol. 1.

I. Ragusa, R. B. Green, *Meditations in the Life of Christ*, 1961, xxiii n. 5; O. Pächt, 'Review', *Medium Aevum*, XXXII (1963), 235; *id.*, 'Kunstlerische Originalität und ikonographische Erneuerung', *Stil und Überlieferung in der Kunst des Abendlandes, Akten des 21 Internationalen Kongresses für Kunstgeschichte, Bonn, 1964*, III, 1967, 264–5, Taf. III, 2, 3, Taf. IV, 1 (reprinted *id.*, *Methodisches zur Kuntsthistorischen Praxis*, 1977).

909. **New College 155** 1379 Padua

Giovanni di Conversino, Dialogus inter Johannem et Literam (250 × 195)

Fine border, historiated initial. In the early style of Cristoforo Cortese. Penwork initials. Cf. nos. 910, 972. [Illus.]

Prov. The dedication copy written in 1379 in the form of a letter by Giovanni di Conversino to Tommaso da Frignano, created cardinal of St. Nereus in 1378. Given to Coluccio Salutai by the author in 1381, fol. 36. Andrew Holes, fellow, archdeacon of Wells (d. 1470).

Emden, 950; B. L. Ullman, *The Humanism of Colluccio Salutati*, 1963, no. 108; E. G. Kohl, 'The works of

Giovanni di Conversino da Ravenna', *Traditio*, XXI (1975), 364.

910. **Keble College 59** Saec. XIV ex. Padua (?)

Psalter with Votive Masses (179 × 127)

Fine borders, historiated and other initials. By Cristoforo Cortese. Good penwork initials. Cf. nos. 909, 972. [Illus.]

Prov. Calendar points to Padua. Sir Thomas Brooke, Huddersfield (d. 1908). Bequeathed by his brother, Canon C. E. Brooke (d. 1911).

C. Huter, 'Cristoforo Cortese in the Bodleian Library', *Apollo*, CXI no. 215 (1980), 17.

911. **New College 89** Saec. XIV ex.
John of Vincelles, Misc. pieces. Sermons (300 × 210)

Good and other initials.

912. **New College 265** Saec. XIV ex. Central. Florence (?)

Boethius, de consolatione philosophiae (252 × 180)

Fine and other initials. [Illus.]

Prov. Annotations by Andrew Holes, fellow, archdeacon of Wells (d. 1470), who was in Florence *c*. 1439–44, and by Thomas Candour in Florence *c*. 1442.

Hunt, *Medieval library*, 326; A. C. de la Mare, 'Humanistic hands in England', Exh., *Manuscripts at Oxford*, XXII, 4.

913. **Keble College 53** Saec. XIV^{2}
Martyrology. Cistercian use. Rule of St. Benedict (268 × 188)

Good initial. Good penwork initials. Added or repainted in a MS. of saec. XIII med.

Blind-stamped calf binding over boards, brass central bosses and corner pieces (four lost), saec. XVI.

Prov. Possibly San Galgano, Tuscany, or one of the daughter houses (marginal note referring to St. Galganus of Siena, fol. 110^{v}). Datable by the latest additions incorporated into the text, *c*. 1246–56. Sir Thomas Brooke, Huddersfield (d. 1908). Bequeathed by his brother, Canon C. E. Brooke (d. 1911).

II. AFTER 1400

1. Central Italy: Florence, Tuscany, Umbria, etc.

914. **All Souls College 94** *c*. 1400 Florence

Coluccio Salutati, Invectiva in Luscum (252 × 152)

Fine border, initial. Penwork initials by the flourisher of Florence, Laur. San Marco 654. Possibly written by the scribe of Laur. 78.2 and Strozzi 95. [Illus.]

Prov. Arms, fol. 3, partly defaced and overpainted.

B. L. Ullman, *The Humanism of Coluccio Salutati*, 1963, 34, 270.

915. **New College 271,** fols. 3–156 *c*. 1400 Florence

Virgil, Aeneid (287 × 210)

Good initial. Other initials added in England mostly removed (cf. no. 609).

Prov. Unidentified English coat of arms, fol. 3: quartered, 1 and 4 gules a saltire argent with a border engrailed azure, 2 and 3 gules three stars of David or with a bordure engrailed azure, with inescutcheon per fess or and gules a lion rampant sable and argent. William Brygon, canon of Salisbury (d. 1469), fol. 1^{v}. Given by John Russel, bp. of Lincoln 1482–94, fol. 1^{v}.

Emden, 1610; Weiss, *Humanism*, 177; Exh., *Duke Humfrey*, 32, no. 59, pl. XXIVa; Hunt, *Medieval library*, 328.

916. **Taylor Institution 8° It. 2** *c*. 1420–40 Florence

Leonardo Bruni, Historia fiorentina. Books I–VI (300 × 223)

Fine and other initials. Humanistic script.

Contemporary binding of stamped leather on pasteboards, Italian. Traces of a fore-edge pattern.

Prov. 'St. Joannis Flisei de Canevali di Christophori', and 'Dico quod non est Christophori sed (?) eius fratris ac amicorum', fol. 1, saec. XVI. 'K. D. Naples 26 June 1733', last pastedown.

917. **St. John's College 116** Saec. XV$^{2}_{4}$ Florence

Seneca, Letters (235 × 158)

Good historiated (one) and other initials, good penwork initials. Added to an earlier MS: 'inceptus scribi 19 Aug. 1385', fol. 8. [Illus.]

Prov. Hicson Vesus, fol. i, saec. XVI.

918. **Corpus Christi College 91** Saec. XV$^{2}_{4}$ Florence

Nonius Marcellus, de proprietate sermonum (254 × 180)

Good initials.

919. **Lincoln College Lat. 20** *c*. 1425–35 Florence

Leonardo Bruni, Misc. works (263 × 183)

Fine initials. By the same hand as Bodl. MS. Buchanan c.1 (cf. Pächt, Alexander, 2, no. 230). Humanistic script.

Prov. Notes in later Italian hands and an erased *ex libris with monogram 'A + B', saec.* XVI, *fol. 84ᵛ*.

de la Mare, *Vespasiano da Bisticci*, 15, pl. IV.

920. **New College 277–9** *c.* 1430–50 Florence

Livy, Decades I, III, IV (prefaced by Boccaccio's Life of Livy) (350 × 250)

Penwork border. Fine and other initials. Corrected and annotated by the English humanistic scribe Thomas Candour. Cf. no. 575. For other MSS. see Index [Illus.]

Prov. William Say, fellow, dean of St. Paul's, London (d. 1468), who may have acquired this set of Livy from Thomas Candour.

Emden, 1650, App. 2158; Exh., *Duke Humfrey*, no. 32; A. C. de la Mare, 'Florentine manuscripts of Livy in the fifteenth century', *Livy, ed.* T. A. Dorey (Greek and Latin Studies, Classical Literature and its influence), 1971, 177–9; Exh., *The Survival of Ancient Literature*, Bodleian Library, Oxford, 1975, no. 153; Hunt, *Medieval library*, 327.

921. **New College 201** *c.* 1440 Florence

Dominicus de Sancto Geminiano, Comm. on Sext. Part II. On paper (425 × 285)

Good border, initial. Penwork initials.

Prov. Arms of Andrew Holes, archdeacon of Wells (d. 1470), fol. 1.

Emden, 950; Exh., *Duke Humfrey*, no. 31, pl. XXIVe.

922. **Magdalen College Lat. 59** *c.* 1440 Florence

Ambrose, Epistola ad Gratianum. Athanasius, contra gentiles (272 × 192)

Good initials (mutilated). Humanistic script.

923. **Magdalen College Lat. 135** *c.* 1440 Florence

Thomas Aquinas, Catena on Matthew and Mark (390 × 260)

Fine border, historiated and other initials (mutilated). [Illus.]

Contemporary Florentine binding of blind-stamped leather on boards.

Prov. Probably owned by Andrew Holes, archdeacon of Wells (d. 1470), who was in Florence *c.* 1439–44.

Exh., *Duke Humfrey*, no. 31.

924. **Magdalen College Lat. 136** *c.* 1440 Florence

Thomas Aquinas, Catena on Luke and John (380 × 245)

Fine and other initials by two hands, the second working in Paduan style. [Illus.]

Prov. Probably owned by Andrew Holes, archdeacon of Wells (d. 1470).

Exh., *Duke Humfrey*, no. 31.

925. **New College 249** *c.* 1440–50 Florence

Cicero, Orations (330 × 240)

Fine initials by an Italian artist active in Florence *c.* 1425–45. Fols. 2–91ᵛ written by Antonio di Mario, his initials 'A.M.', fol. 66, cf. nos. 934, 943, 946. Fols. 92–378 written by Dominicus Cassii de Narnia working in Florence *c.* 1442–59, cf. no. 932. [Illus.]

Prov. Given by John Baker, fellow (d. 1488), executor of Andrew Holes, the original owner (d. 1470), fol. iiᵛ.

M. R. James, *The Chaundler MSS.* (Roxburghe Club), 1916, 5; J. W. Bennett, 'Andrew Holes . . .', *Speculum*, XIX (1944), 324; B. L. Ullman, *The origin and development of humanistic script*, 1960, 105; Emden, 93, 950; R. W. Hunt, 'Humanistic script in Florence in the early fifteenth century', *Calligraphy and Palaeography. Essays presented to Alfred Fairbank, ed.* A. S. Osley, 1965, 274, pl. 48b; B. L. Ullman, 'More Humanistic Manuscripts', *ib.*, 51; Weiss, *Humanism*, 79 n. 3; Exh., *Duke Humfrey*, no. 29.

926. **Magdalen College Lat. 37, part I,** fols. 1–60ᵛ *c.* 1442 Florence

Plutarch, Life of Marius, etc. Latin translation by Antonio Pacini, with dedication to Duke Humfrey (268 × 172)

Good initials. Cf. no. 995.

Prov. Presumably the presentation copy to Humfrey, duke of Gloucester, his *ex libris* (erased), fol. 60ᵛ.

Exh., *Duke Humfrey*, no. 13A; Weiss, *Humanism*, xiv, 52 n. 7, 64 n. 8.

927. **New College 63,** fols. 123–289 Saec. XV$^{2}_{4}$ (before 1444) Florence

Thomas Aquinas, Comm. on St. Paul's Epistles (360 × 240)

Fine historiated and other initials. Close in style to Zanobi Strozzi (cf. Bodl. MS. Canon. Misc. 400, Pächt, Alexander, 2, no. 236, pl. XXII). Fine and other penwork initials. Added to a MS. written in Padua, cf. no. 981. [Illus.]

Prov. Annotated by Andrew Holes, fellow, King's proctor at the papal court, archdeacon of Wells (d. 1470) who left Florence in 1444.

928. **Balliol College 35B** *c.* 1444–53 Florence

Robert Grosseteste, Comm. on the Gospels (incomplete) (390 × 260)

Good initials. Penwork initials. In the same style as Balliol MS. 126 (no. 935). Added in Florence to a MS. written for William Gray by a Dutch scribe Tielmann Reynerszoon in 1443, possibly at Cologne, colophon fol. 170ᵛ. Marginal drawing with his name, fol. 100ᵛ. Cf. nos. 443, 511.

Prov. William Gray, bp. of Ely 1454–78, fol. iᵛ.

Mynors, *Catalogue*, xxvi; Ker, *College Libraries*, 309 n. 81.

929. **Balliol College 262** *c.* 1444–53 Florence

Nonius Marcellus, de proprietate sermonum (205 × 196)

Good border (mutilated). Good and other initials (mutilated). Humanistic script.

Prov. William Gray, bp. of Ely 1454–78, fol. 1ᵛ.

930. **Balliol College 248A** *c.* 1444–53 Florence

Cicero, Orations I (incomplete) (362 × 240)

Fine initials (mutilated). By two hands, one identifiable in MSS. Balliol 135, 242 (nos. 944, 940). Humanistic script.

Prov. William Gray, bp. of Ely 1454–78.

931. **Balliol College 248B** *c.* 1444–53 Florence

Cicero, Orations II (incomplete) (362 × 240)

Fine initials. Related to the style of Filippo Torelli. Cf. nos. 936, 937. Humanistic script by Gherardo di Giovanni del Ciriagio, citizen and notary of Florence, colophon fol. 225, Cf. no. 955. [Illus.]

Prov. William Gray, bp. of Ely 1454–78, fol. 1ᵛ.

A. C. de la Mare, 'Vespasiano da Bisticci and Gray', *J.W.C.I.* xx (1957), 174 n. 12; B. L. Ullman, *The origin and development of humanistic script*, 1960, 112, 117.

932. **Balliol College 248C** *c.* 1444–53 Florence

Cicero, Letters (362 × 240)

Fine border (mutilated), initials. Humanistic script attributable to Dominicus Cassii de Narnia working in Florence from *c.* 1442 to 1459. Cf. no. 925.

Prov. William Gray, bp. of Ely 1454–78, fol. 1ᵛ.

933. **Balliol College 248D** *c.* 1444–53 Florence

Cicero, de officiis, etc. (350 × 240)

Fine initials (mutilated) in the style of Joachinus de Gigantibus, cf. nos. 938, 942, 943, 945, 946. Written by the same hand as MS. Balliol 138 (no. 942).

Prov. William Gray, bp. of Ely 1454–78, front pastedown.

934. **Balliol College 248E** 1445 Florence

Cicero, Rhetorical works (413 × 245)

Fine borders, initials (mutilated). Humanistic script by Antonio di Mario, citizen and notary of Florence, colophon fol. 251. Cf. nos. 925, 943, 946.

Prov. William Gray, bp. of Ely 1454–78, fol. 1ᵛ.

A. C. de la Mare, 'Vespasiano da Bisticci and Gray', *J.W.C.I.* xx (1957), 174 n. 12; B. L. Ullman, *The origin and development of humanistic script*, 1960, 103, 105, 108; Exh., *Duke Humfrey*, 28.

935. **Balliol College 126** *c.* 1444–53 Florence

Petrarch, Liber rerum familiarium (incomplete (465 × 310)

Fine initials. Penwork initials. Similar in style to Balliol MS. 35B (no. 928). Added to a MS. written in humanistic script probably in Oxford *c.* 1442 by a scribe who copied several MSS. for William Gray, cf. nos. 438, 442, 444.

Prov. William Gray, bp. of Ely 1454–78, parchment label with his arms from the old cover pasted on fol 1ᵛ.

Exh., *Duke Humfrey*, 26; N. Mann, 'Petrarch's Role in Humanism', *Apollo*, XCIV no. 115 (1971), 180, fig. 4; *id.*, 'Petrarch manuscripts in the British Isles', *Italia medioevale e umanistica*, XVIII (1975), no. 153.

936. **Balliol College 140** *c.* 1444–53 Florence

Virgil, Works (358 × 210)

Fine initials (mutilated) by Filippo Torelli, cf. nos. 931, 937. Humanistic script by the scribe of Venice, Bibl. Marc. Lat. XII.68 and Brit Libr. Add. 24638. [Illus.]

Prov. William Gray, bp. of Ely 1454–78, fol. 2ᵛ. Thomas Holland, fellow, chaplain of the College 1579, fol. 31.

937. **Balliol College 249** *c.* 1444–53 Florence

Pliny the elder, Natural History, etc. (390 × 265)

Fine border (mutilated), fine and other initials. In the early style of Filippo Torelli, cf. nos. 931, 936. Humanistic script attributable to the early hand of Piero Strozzi. [Illus.]

Prov. William Gray, bp. of Ely 1454–78, fol. 1^v.

A. C. de la Mare, 'Messer Piero Strozzi', *Calligraphy and Palaeography. Essays presented to Alfred Fairbank, ed.* A. S. Osley, 1965, 60, no. 22.

938. **Balliol College 237** *c.* 1444–53 Florence

Paulus Diaconus, Epitome of Festus (376 × 240)

Fine border (mutilated), initials (unfinished and mutilated) in the early style of Joachinus de Gigantibus, cf. nos. 933, 942, 943, 945, 946. Humanistic script attributable to Theodericus Werken, cf. no. 447. For other MSS. see Index. [Illus.]

Prov. William Gray, bp. of Ely 1454–78, fol. 1^v.

de la Mare, *A fragment*, 285 n. 3.

939. **Balliol College 128** *c.* 1444–53 Florence or Rome

Georgius Trapezuntius, Comm. on Cicero's pro Ligario (244 × 112)

Border, initial (rubbed and mutilated). Humanistic script by the scribe of MS. Balliol 131^3. Cf. no. 562.

Prov. William Gray, bp. of Ely 1454–78.

940. **Balliol College 242** *c.* 1444–53 Florence

Aristotle, Ethics and Politics, (incomplete). Leonardo Bruni's translation (357 × 253)

Good initials by the same illuminator as MSS. Balliol 135, 248A. (Cf. nos. 944, 930).

Prov. William Gray, bp. of Ely 1454–78, offset of his arms, fol. 1^v.

941. **Balliol College 139** *c.* 1444–53 Florence

Quintilian, Declamationes (incomplete) (250 × 175)

Good and other initials. Humanistic script, written probably at Florence by the scribe of MS. Balliol 233 (cf. no. 945).

Prov. Fly-leaf lost but probabily William Gray, bp. of Ely 1454–78.

942. **Balliol College 138** *c.* 1444–53 Florence

Quintilian, Institutio oratoria (incomplete) 348 × 233

Good initials possibly by Joachinus de Gigantibus, cf. nos. 933, 938, 943, 945, 946. Written in Florence by the scribe of Balliol MS. 248D (no. 933).

Prov. Label from the old cover with Gray's arms, fol. ii^v.

943. **Balliol College 154** 1447 Florence

John Chrysostom, Letters (incomplete) (370 × 250)

Fine and other initials (mutilated) mostly by Joachinus de Gigantibus, cf. nos 933, 938, 942, 945, 946. Written for William Gray in humanistic script by Antonio di Mario, Florence, 1447, colophon fol. 266, cf. nos. 925, 934, 946. 'Lege feliciter, mi suavissime Ghuiglelme', fol. 2^v. [Illus.]

Prov. William Gray, bp. of Ely 1454–78, fol. 2^v.

Exh., *Italian Illuminated MSS.*, 1948, no. 91; B. L. Ullman, *The origin and development of humanistic script*, 1960, 103, no. 32, 105, 107, 108; Exh., *Duke Humfrey*, 28.

944. **Balliol College 135** *c.* 1447–50 Florence

Guarino of Verona, Speeches. Letters (350 × 220)

Fine borders, initials (mutilated). By the same illuminator as Balliol MSS. 242 and 248A (nos. 940, 930). Humanistic script attributable to Theodericus Werken, cf. no. 447. For other MSS. see Index.

Prov. Fly-leaf missing but no doubt given by William Gray, bp. of Ely 1454–78. John Mathewes, 1644, fol. 60.

Mynors, *Catalogue*, xxxii; P. T. Eden, 'William Gray, bishop of Ely and three Oxford MSS of Seneca', *Classica et Medievalia*, XXI (1960), 40; de la Mare, *A fragment*, 285 n. 3.

945. **Balliol College 233** *c.* 1447–53 Florence

Lorenzo Valla, Elegantiae, etc. (360 × 245)

Fine border, fine and other initials (mutilated). By Joachinus de Gigantibus, cf. nos. 933, 938, 942, 943, 946. Written in humanistic script after March 1447 (reference to Nicholas V) and before 1453, Gray's departure from Italy.

946. **Balliol College 78B** 1448 Florence

Johannes Climacus, Scala spiritualis, etc. (372 × 255)

Fine border (mutilated), initials, in the early style of Joachinus de Gigantibus, cf. nos. 933, 938, 942, 943, 945. Written for William Gray in Florence in humanistic script by Antonio di Mario, colophon fol. 170^v. Cf. nos. 925, 934, 943. [Illus.]

Prov. William Gray, bp. of Ely 1454–78, fol. 3.

B. L. Ullman, *The origin and development of humanistic script*, 1960, 103; Exh., *Duke Humfrey*, no. 44.

947. **Taylor Institution 8° It. 3** Saec. xv^1 Florence

Dante, Paradiso with comm. by Francesco da Buti. In Italian. On paper (290 × 210)

Fine historiated and other initials.

Prov. Stamp of the Salviati (?) family, fol. i. Arms of Pope Pius VI (d. 1799) on the binding. Bought at Naples in 1862 by Ed. Giannini, Pisa.

E. Moore, *Contributions to the textual criticism of the Divina Commedia*, 1889, 549–50.

948. **Lincoln College Lat. 21** Saec. xv med. Florence

Aristotle, Ethics. Leonardo Bruni's translation (235 × 160)

Fine initials. Humanistic script. [Illus.]

Prov. Robert Flemmyng, dean of Lincoln (d. 1483).

Weiss, *Catalogues*, 349 (69) n. 90; de la Mare, *Vespasiano da Bisticci*, 15.

949. **Lincoln College Lat. 37** *c.* 1450 Florence

St. Cyril. Comm. on Genesis (217 × 145)

Good and other initials. Humanistic script.

Prov. Notes by Thomas Gascoigne, chancellor of Oxford University (d. 1458), fol. 1v. Robert Flemmyng, dean of Lincoln (d. 1483), fol. ii.

Emden, 700, 746; de la Mare, *Vespasiano da Bisticci*, 16.

950. **Lincoln College Lat. 39** *c.* 1450 Florence

Cicero, Orations (312 × 222)

Fine border (mutilated), initials. Semi-humanistic script. [Illus.]

Prov. Robert Flemmyng, dean of Lincoln (d. 1483).

Weiss, *Catalogues*, 349 (59) n. 96; Emden, 700; de la Mare, *Vespasiano da Bisticci*, 16.

951. **Lincoln College Lat. 42** *c.* 1450 Florence

Cicero, Epistola ad Brutum (incomplete), etc. (260 × 190)

Good initials (mutilated). Humanistic script.

Prov. Robert Flemmyng, dean of Lincoln (d. 1483), fol. i.

de la Mare, *Vespasiano da Bisticci*, 15.

952. **Lincoln College Lat. 49** *c.* 1450 Florence

Dionysius Areopagita, de caelesti hierarchia (260 × 180)

Fine border (mutilated), initials. Humanistic script.

Prov. Robert Flemmyng, dean of Lincoln (d. 1483).

de la Mare, *Vespasiano da Bisticci*, 16.

953. **Lincoln College Lat. 59** *c.* 1450 Florence

Aulus Gellius, Noctes atticae (335 × 228)

Fine border (mutilated), initials. Humanistic script. [Illus.]

Prov. Robert Flemmyng, dean of Lincoln (d. 1483), fol. iv.

Emden, 700; de la Mare, *Vespasiano da Bisticci*, 16; Weiss, *Humanism*, 104 n. 8.

954. **Taylor Institution 8° It. 4,** fols. 1–41 *c.* 1450–60 Florence

Petrarch, Trionfi. In Italian (215 × 125)

Good initial. Humanistic script.

Prov. 'Questo libro a di Lorenzo di Luca Ubertini ...', saec. xv/xvi, written over an earlier inscription 'Questo libro sia di Tomas di Filippo', saec. xv ex., fol. i.

N. Mann, 'Petrarch manuscripts in the British Isles', *Italia medioevale e umanistica*, xviii (1975), no. 249.

955. **Corpus Christi College 84** *c.* 1450–60 Florence

Ps.-Pliny, de viris illustribus (225 × 150)

Good border (unfinished), initial. Humanistic script. Rubrics by the Florentine scribe Gherardo di Giovanni del Ciriagio, fols. 47, 48. Cf. no. 931. [Illus.]

Prov. Bought in London in 1464 by John Shirwood, bp. of Durham 1484–93, his *ex libris*, 1491, fol. i. Given by the founder, Richard Foxe, bp. of Winchester 1501–28, fol. 1.

Emden, 717, 1693; F. Saxl. R. Wittkover, *British Art and the Mediterranean*, 1948, pl. 36a; Exh., *Duke Humfrey*, no. 101.

956. **University College 54** *c.* 1455–60 Florence

John Chrysostom, Homilies (330 × 215)

Fine border (mutilated), historiated and other initials. By the same hand as Bodl. MS. Digby 231, Francesco d'Antonio del Cherico. Humanistic script. [Illus.]

Prov. Thomas Clifford, Franc. Clethero, saec. xv (?), fol. 227.

957. **Lincoln College Lat. 47** *c.* 1460 Florence

Cyprian, Epistola ad Donatum, etc. (320 × 210)

Fine border (mutilated), initials. Humanistic script. 'Vespasianus librarius Florentinus hunc librum Florentie transcribendum curavit', fol. iv. [Illus.]

Prov. Given after 1474 by Robert Flemmyng, dean of Lincoln (d. 1483).

de la Mare, *Vespasiano da Bisticci*, 12, 14, pls. i, v.

958. **Lincoln College Lat. 45** *c.* 1460
Florence

Donatus, Comm. on Terence (315 × 210)

Fine border (mutilated), initials. Humanistic script by the scribe of Lincoln College Lat. 44 and Edinburgh Univ. 200. Signed by the scribe 'Vespasianus librarius fecit scribi Florentie', fol. i^v.

Prov. Note by Robert Flemmyng, dean of Lincoln (d. 1483), fol. II.

Weiss, *Catalogues*, 350 (75) n. 96; Emden, 700; de la Mare, *Vespasiano da Bisticci*, 12–14, pl. II; Exh., *Duke Humfrey*, no. 64.

959. **Lincoln College Lat. 44** *c.* 1460
Florence

Donatus, Comm. on Vergil (305 × 223)

Good border (mutilated), initials. Humanistic script by the scribe of Lincoln College Lat. 45 and Edinburgh Univ. MS. 200.

Prov. Given by Robert Flemmyng, dean of Lincoln (d. 1483), fol. 1^v.

Weiss, *Catalogues*, 350 (88), n. 104; Emden, 700; de la Mare, *Vespasiano da Bisticci*, 13, 14, pl. III; Exh., *Duke Humfrey*, no. 64.

960. **Merton College 326** (H.2.10) *c.* 1465–75
Florence

Petrarch, Trionfi, etc. In Italian (239 × 150)

Good border (mutilated), initials. Humanistic script by Giovanni Francesco Marzi di San Giminiano.

Prov. Samuel Stokes, fol. 1, saec. XVI–XVII.

N. Mann, 'Petrarch manuscripts in the British Isles', *Italia medioevale e umanistica*, XVIII (1975), no. 244.

961. **Lady Margaret Hall, Borough 1,** fols $84–246^v$
c. 1480–90
Central Italy (?)

Book of Hours (188 × 132)

Good historiated initials. Additions to a MS. of *c.* 1460–70, cf. no. 997. [Illus.]

Prov. Cf. no. 997.

962. **Keble College 60, 61, 62** Saec. XV ex.
Florence

Book of Hours. Use of Rome (118 × 78)

Fine miniatures, historiated borders, historiated and other initials. By the workshop of Attavante. Originally one manuscript. Retouched by Caleb Wing *c.* 1850. Written in humanistic script by Alessandro Verrazzano. [Illus.]

Prov. John Boykett Jarman sale, Sotheby's, 13 June 1864, lot 15. Sir Thomas Brooke, Huddersfield (d. 1908). Bequeathed by his brother Canon C. E. Brooke.

963. **Keble College 27** 1516
Spoleto

Missal (395 × 278)

Fine miniatures, borders in two styles, the second on inserted leaves. Fine historiated and other initials. [Illus.]

Prov. Written for Francisco Eroli, bp. of Spoleto (1497–1540): frontispiece with his arms and date, fol. 1^v, arms repeated in borders, *passim.* Sir Thomas Brooke, Huddersfield (d. 1908). Bequeathed by his brother, Canon C. E. Brooke (d. 1911), first pastedown.

2. South Italy: Rome, Naples, etc.

964. **Keble College 30** 1404
Monte Cassino

Breviary. Use of Monte Cassino (287 × 205)

Fine historiated and other borders, historiated and other initials. Written by Francesco Guantari from Naples, 1404, colophon fol. ii. [Illus.]

Blind-stamped red leather binding on boards, two leather straps (one lost) with metal pins on back cover, Italian, saec. XVI in.

Prov. Written for abbot Enrico Tomacelli of Monte Cassino, represented on fols. 9, 89, and presumably, fols. 127^v, 146, 360^v, his arms in the initial on fol. 15. Arms overpainted on fols. 9, 127^v, 146, 229, are of the Petrucci family of Rome. The original colophon, now bound before the MS., is copied in its correct place, fol. 524, by a different hand. Sir Thomas Brooke, Huddersfield (d. 1908). Bequeathed by his brother Canon C. E. Brooke (d. 1911).

Exh., *Italian Illuminated MSS.*, 1948, no. 58.

965. **Magdalen College Lat. 76** Saec. XV^2_4
Rome (?)

Life of Gregory Nazianzus, John Chrysostom, etc. (237 × 192)

Fine initials. Good penwork initials. Written in humanistic script by Johannes Rodenburg (colophon fol. 90^v), who also wrote and signed MS. Vat. lat. 1801 for Pope Nicolas V in 1452 and MS. Vat. lat. 1812. [Illus.]

966. **New College 250** Saec. XV med.
Rome (?)

Cicero, de oratore, etc. (345 × 240)

Fine initials. Written in two humanistic hands: the first probably Italian, fols. $1–41^v$, l. 32, the second that of an English scribe, Thomas Candour, fols. 41^v, l. $32–98^v$, cf. no. 575. For other MSS. see Index.

Prov. William Say, fellow, dean of St. Paul's, London (d. 1468), fol. i^{v}.

Emden, 1650, App. 2158; Exh. *Duke Humfrey*, no. 33, pl. IX.

967. **St. John's College 131** Saec. xv$^{3}_{4}$ Naples

Psalter, Offices of the Virgin, etc. Use of the Order of St. John of Jerusalem (220 × 152)

Fine borders, historiated and other initials. Penwork initials. By Cola Rapicano (d. 1488). [Illus.]

Prov. Calendar with obits and dates relating to the Weston family of Buxton Overy, Leicestershire, 1558–82. Obit of Edward Boyd, 1592. Sir William Paddy, commoner, physician to James I (d. 1634), fol. 2.

3. North-East Italy: Venice, Padua, etc.

968. **Balliol College 185** *c.* 1400 North-east Italy

Agostino Favaroni, Comm. on the Apocalypse, etc. On paper (209 × 215)

Initials.

Prov. William Gray, bp. of Ely 1454–78.

969. **Magdalen College Lat. 182**, fols. 103–123^{v} Saec. xv in. Padua

De operatione astrolabi and other treatises (330 × 225)

Fine initials. Penwork initials, diagrams. Bound together with astronomical tables written by Thomas Corry, monk of Muchelney, 1440.

Medieval binding of blind-stamped leather on boards.

Prov. Given by John Bolderne in 1577, fol. 1.

Ker, *Pastedowns*, 113, no. 1219.

970. **Balliol College 288** Saec. xv in. North-east Italy. Padua (?)

Giovanni di Conversino, Various works, I (282 × 207)

Good border, historiated and other initials. Companion volume with Paris, Bibl. Nat. Lat. 6494, written by the same Flemish scribe, Jacobus Witte (?). [Illus.]

Prov. Probably acquired by William Gray, bp. of Ely 1454–78, in Padua *c.* 1445.

Mynors, *Catalogue*, xxxii, 309; B. G. Kohl, 'The works of Giovanni di Conversino da Ravenna: a Catalogue of Manuscripts and editions', *Traditio*, XXXI (1975), 352.

971. *Vacat*

972. **Keble College 31** *c.* 1420 Venice

Breviary (incomplete). Franciscan use (127 × 96)

Good historiated and other initials. By Cristoforo Cortese. Cf. nos. 909, 910. [Illus.]

Prov. Made for a member of one of the Franciscan orders (liturgical evidence). Sir Thomas Brooke, Huddersfield (d. 1908). Bequeathed by his brother Canon C. E. Brooke (d. 1911).

C. Huter, 'Cristoforo Cortese in the Bodleian Library', *Apollo*, CXI no. 215 (1980), 16–7.

973. **Lincoln College Lat. 40** Saec. xv$^{2}_{4}$ Venice or Padua

Cicero, Orations. On paper (305 × 223)

One fine and other initials. Humanistic script. Watermarks suggest North-east Italy, saec. xv$^{1}_{4}$. [Illus.]

Prov. Given by Robert Flemmyng, dean of Lincoln (d. 1483), fol. 1.

Emden, 700.

974. **St. John's College 192** Saec. xv$^{2}_{4}$ North-east Italy

Juvenal, Satires (165 × 105)

Initials (unfinished). Humanistic script.

Prov. Ben Jonson, dramatist (d. 1637), fol. 1.

975. **Magdalen College Lat. 184** *c.* 1440–60 Ferrara

Eusebius, Chronicon (310 × 215)

Fine borders, historiated and other initials. Humanistic script. [Illus.]

Notes and Queries, 6 ser. VII (1883), 422.

976. **Balliol College 137** *c.* 1444–53 Ferrara (?)

Lysias, Orations. Translation by Francesco Filelfo (220 × 150)

Fine initials. Humanistic script by 'Ang. Nik', fol. 20^{v}.

Prov. Probably William Gray, bp. of Ely 1454–78, cf. Mynors, *Cat.*

Exh., *Italian Illuminated MSS.*, 1948, no. 90.

977. **Balliol College 132** *c.* 1444–53 (?) North-east Italy

Gasparinus Barzizius, Exordia, etc. (incomplete) (302 × 225)

Good initials, by the same hand as Bodl. MS. Canon. Class. Lat. 302. Humanistic script. '. . . scripsit Padue Damianus de Pola, 1425' (?), fol. 60^{v}, but probably copied from an exemplar (cf. Mynors, *Catalogue*).

Prov. William Gray, bp. of Ely 1454–78, fol. ii.

Exh., *Italian Illuminated MSS.*, 1948, no. 89; Pächt, Alexander, 2, no. 875.

978. **Balliol College 295** 1445
Padua (?)

Sicco Polentone and others. Comm. on Cicero. On paper (330 × 287)

Good border (mutilated). Good initials in two styles. Written in humanistic script for William Gray in 1445 by Theodericus Werken, probably in Padua, colophon fol. 122^v, cf. no. 447. For other MSS. see Index.

Prov. William Gray, bp. of Ely 1454–78. Annotations by Richard Bole (d. 1477). John Braxton, scholar of the College 1581, fols. 94, 95.

Weiss, *Humanism*, 86, n. 2; Mynors, *Werken*, 99, 104; Exh., *Duke Humfrey*, no. 43; de la Mare, *A fragment*, 285, ns. 3, 5.

979. **Lincoln College Lat. 43** *c.* 1446
Padua

Cicero, de officiis (144 × 104)

Good initials. Written in semi-humanistic script by Robert Flemmyng, signed on the end fly-leaf with, above, a note 'V(isto?) per me ss.o(?) 10 ottobr. 1446' in an Italian hand.

Prov. Robert Flemmyng, dean of Lincoln (d. 1483).

Emden, 700; de la Mare, *Vespasiano da Bisticci*, 11 n. 1; Exh., *Duke Humfrey*, no. 60.

980. **Keble College 3** 1447
North-east Italy

Alexander de Villa Dei, Doctrinale (276 × 190)

Border, initial. Penwork initials. Colophon fol. 40^v: 'Deo gratias Amen. 1447'.

Prov. Given by Rev. Greville J. Chester in 1882, first pastedown.

981. **New College 63,** fols. 1–120^v Saec. xv^1
Padua

Thomas Aquinas, Comm. on St. Paul's Epistles (360 × 240)

Fine border, historiated (defaced) and other initials. Penwork initials. Cf. no. 927.

Prov. Arms (defaced), fol. 1, possibly of Biagio Molin, bp. of Grado 1427–34 (translated to Jerusalem). Annotated by Andrew Holes who was in Padua in 1439.

982. **Magdalen College Lat. 83** Saec. xv^1
North-east Italy

Cicero, Epistolae ad familiares (282 × 185)

Good historiated initial. Good penwork initials. Humanistic script. [Illus.]

983. **Lincoln College Lat. 41** *c.* 1450
Padua (?)

Cicero, Orations (245 × 155)

Good initial. Humanistic script.

Prov. Annotated by Robert Flemmyng, dean of Lincoln (d. 1483). Given by him to the College after 1474, fol. 249.

Emden, 700; de la Mare, *Vespasiano da Bisticci*, 11 n. 1; Exh., *Duke Humfrey*, 36.

984. **Magdalen College Lat. 64** Saec. xv med.
Padua

Servius, Ars grammatica (232 × 150)

Good initial, other initials (unfinished). Humanistic script.

Exh., *Duke Humfrey*, no. 81.

985. **All Souls College 73** Saec. xv med.
Padua (?)

Petrus de Azzalata, de chirurgia. On paper (390 × 278)

Good initial (rubbed). Penwork initials of English type. Written by an Englishman possibly studying at Padua.

986. **Christ Church 113–14** 1456
Ferrara (?)

Virgil, Works (254 × 170)

Fine border, fine and other initials. Written in humanistic script by Matteo Contugi of Volterra, colophon fol. 156^v. [Illus.].

Prov. Arms: barry or and sable on a chief gules a rose argent in MS. 113, fol. 2. Given in 1765 by John Moore, later archbp. of Canterbury, 1783–1805.

987. **Corpus Christi College 79** 1458
Ferrara

Livy, Decade I (225 × 150)

Good border, initial. Humanistic script by Antonius F., notary, fol. 163.

Prov. Given by the founder, Richard Foxe, bp. of Winchester 1501–28, fol. 2.

988. **University College 94** *c.* 1460
Ferrara

Diogenes Laertius, Lives. Translation by Ambrogio Traversari (262 × 175)

Fine border, fine and other initials. By the same hand as Bodl. MS. Canon. Class. Lat. 100. Humanistic script. [Illus.]

Contemporary Italian binding of blind-stamped leather on boards.

Prov. Arms, fol. 4, azure a vase argent containing a tree vert, overall a bend or (?). Initials 'I. V.'. Given by John Hales, gentleman commoner, *c.* 1675.

Hunt, *Manuscript*, 21; Pächt, Alexander, 2, no, 406.

989. **Corpus Christi College 66** *c.* 1460
Padua

Manilius, Astronomicon; Comm. on Ovid; Ibis. On paper (335 × 235)

Good initials close in style to Bodl. MS. Auct. F. 1. 13 (cf. Pächt, Alexander, 2, no. 606). Written by the same humanistic scribe who there signs himself 'V.f.I' (cf. also Pächt, Alexander, 3, no. 988). Annotations in a contemporary English humanistic hand (cf. *Duke Humfrey*, nos. 72, 74, etc.). [Illus.]

Contemporary English binding of faded pink leather on boards.

Exh., *Duke Humfrey*, 41–2, no. 78.

990. **Keble College 34** Saec. xv$\frac{3}{4}$
Venice (?)

Modus visitandi domos Ordinis Carthusiensis, etc. (138 × 92)

Good borders, initials.

Binding of gold-tooled brown leather on boards with 'Carthusia' on the front cover, metal clasps (one lost), Italian, saec. XVI in.

Prov. 'Carthusie pafie' (Certosa at Pavia), fol. 1, saec. XVI (?). William Bragge sale, Sotheby, Wilkinson & Hodge's, 7 June 1876, lot. 131. Sir Thomas Brooke, Huddersfield (d. 1908). Bequeathed by his brother, Canon C. E. Brooke (d. 1911).

991. **New College 342** Saec. xv^2
Ferrara (?)

Anonymous, De regimine principum, etc. (130 × 94)

Fine and other initials. Humanistic script. [Illus.]

Contemporary Italian binding of gold-stamped leather on boards, centre medallion with 'JHS' on each cover.

992. **Wadham College 15** (A.5.4) 1523
Venice

Commissio from Doge Andrea Gritti (1523–38) to Bernardino Taiapiero (250 × 170)

Fine historiated borders, arms of Taiapiero..

Written in humanistic script. Fols. 20^v–26 written in Italian by Petrus Grasolarius, notary, 1523. [Illus.]

Prov. Book-plate with arms of Richard Warner of Woodford Row, Essex, commoner of the College, first pastedown; bequeathed in 1775.

993. **Keble College 5** 1536
North-east Italy

Pontificial. Use of Trent (430 × 283)

Good miniatures, borders. In two styles. Good historiated and other initials. [Illus.]

Contemporary gold tooled leather binding, resewn. At foot '1536'.

Prov. Written for Cardinal Bernard Cles, archbp. of Trent (1514–39), his arms, dedication and date on fol. 1^v. His arms also in the borders *passim*. William Bragge sale, Sotheby, Wilkinson and Hodge's, 7 June 1876, lot 381. Bequeathed by Dr. Henry P. Liddon, canon of St. Paul's, London (d. 1890).

4. North-West Italy: Milan, Lombardy, etc.

994. **Christ Church 507** Saec. xv$\frac{2}{4}$
Milan

Ovid, Heroides (62 × 42)

Good historiated initial. Errors in spelling suggest the MS. was possibly written in the Veneto, saec. XIV^2.

Prov. 'Iste liber ovidi est. . . . Petri et Antonii. . . . de Nallate', saec. XV, fol. i^v. Bequeathed by Dr. S. G. Owen (d. 1940).

S. G. Owen, 'A manuscript of Ovid's Heroides', *Classical Quarterly*, XXXI (1937), 1–15.

995. **Magdalen College Lat. 37, part II,** fols. 61–280^v *c.* 1441
Milan

Ptolemy, Cosmography. Latin translation by Jacopo di Angelo da Scarperia (262 × 172)

Good initial. Written in humanistic script, partly by Piercandido Decembrio, for Humfrey, Duke of Gloucester. Cf. no. 926.

Prov. Duke Humfrey's *ex libris* (erased), fol. 279^v. Magister John Manson, saec. XV/XVI, fol. 280^v.

Weiss, *Humanism*, xiv, 52 n. 8, 68 n. 8; Exh., *Duke Humfrey*, 13B.

996. **Mansfield College, s.n.** Saec. XV med.
North-west Italy

Missal (fragment) (280 × 200)

Good initials (rubbed and mutilated). Good penwork initials.

Prov. Given by J. L. Cherry in 1909.

997. **Lady Margaret Hall, Borough 1,** fols. 15–81
c. 1460–70
Lombardy

Book of Hours, Use of Rome (188 × 132)

Fine miniature, borders, historiated initials by the 'Ippolita Master'. Good and other initials. Cf. no. 961. [Illus.]

Prov. Calendar points to a house of Augustinian hermits after 1446 (canonization of Nicholas of Tolentino). 'Finis Venetiis, 1490', fol. 247, probably a much later addition. Caraccioli family, Naples, 1574, fol. 247^{v}. Sale of E. F. Bosanquet, Sotheby's, 24 Jan. 1944, lot 179. Bequeathed by Miss C. M. Borough, 1973.

998. **Campion Hall 4** Saec. xv^{2} Lombardy

Apollinarius Cremonensis, Comm. on Aristotle's Posterior analytics. On paper (406 × 290)

Good border, historiated and other initials. Paper manufactured in Milan, *c.* 1455, 1472–5 (watermarks C. M. Briquet, *Les filigranes*, 1968, III, nos. 6597–99; IV, no. 13048).

Prov. Chiaravalle abbey, nr. Milan, partly erased inscription, saec. xv^{2}, fol. 84. 'A.B.', fol. 94, saec. XVI.

999. **Keble College 63** Saec. xv ex. Milan

Book of Hours. Use of Rome. In Latin and Italian (170 × 125)

Fine miniatures, historiated and other borders in two styles, historiated and other initials. Fine penwork initials. By the Master of the Arcimboldi Missal and associates (cf. Pächt and Alexander, 2, no. 991). [Illus.]

Prov. Written *c.* 1480 for Antonio Landriani, treasurer of Lodovico il Moro, duke of Milan. His arms with monogram 'AN.LA', fol. 14, his portrait, fol. 22, and the name 'Antonio' in the prayers, fols. 134^{v}, 135^{v}. Henry Jadis sale, 19 May 1826, lot 96 (?). Dawson Turner sale, Sotheby's, 7 March 1853, lot 1882. Sir William Tite sale, Sotheby, Wilkinson and Hodge's, 18 May 1874, lot 2183. Sir Thomas Brooke, Huddersfield (d. 1908). Bequeathed by his brother Canon C. E. Brooke (d. 1911).

Exh., *Italian illuminated MSS.*, 1948, no. 74.

5. Italy: Unlocalized

1000. **Magdalen College Lat. 62** Saec. $xv\frac{1}{4}$

Cicero, de divinatione. On paper (247 × 155)

Initials. Pen-flourishes added in England later in saec. XV. Humanistic script.

1001. **New College 256** Saec. xv med.

Guarino da Verona, de dipthongis, etc. (283 × 214)

Good penwork initial in Italian style, fol. 11. Written in humanistic script on Italian parchment by the English scribe Thomas Candour. Cf. no. 575. For other MSS. see Index.

Emden, App. 2158; Exh., *Duke Humfrey*, 32, no. 56.

1002. **New College 286,** fols. 121–25 Saec. $xv\frac{3}{4}$

Lucian, Vita Demonactis. Latin translation by Lapo Castelliunculo. On paper (290 × 215)

Good initial. Humanistic headings.

Prov. Annotations by James Goldwell, fellow of All Souls College, Oxford, bp. of Norwich 1472–1499.

SPANISH MANUSCRIPTS

This is another small section. The Bandini Encyclopedia, Balliol College MS. 238 E (no. 1004), is not only interesting for its iconography, but also because it is a late example of a scribe, Laurentius Dyamas, who also executed the illumination in the form of pen drawings.

1003. **New College 255** Saec. XIII ex.
Guido Faba, Rota nova (265 × 177)

Good miniatures, historiated and other initials. Marginal drawing, fol. 46v. [Illus.]

Prov. Perhaps in England before 1350 (cf. Denholm-Young).

N. Denholm-Young, 'The Cursus in England', *Oxford Essays in Medieval History presented to H. E. Salter*, Introduction by F. M. Powicke, 1934, 94 f.; E. H. Kantorowicz, 'An "Autobiography" of Guido Faba', *Medieval and Renaissance Studies*, I (1941–43), 253–80; T. Kurose, *Miniatures of the Goddess Fortune in Mediaeval Manuscripts*, 1977, pl. 141.

1004. **Balliol College 238E** *c.* 1445
Spanish (?)
Domenico Bandini, Fons memorabilium universi, V (357 × 246)

Fine marginal drawings by the scribe, Laurentius Dyamas (signed, fol. 162), who wrote and illuminated Bodl. MS. Douce 204 for Guillaume de Montjoie, bp. of Béziers, 1424–51 (cf. Pächt, Alexander). Cf. nos. 865, 867. [Illus.]

Prov. William Gray, bp. of Ely 1454–78, fol. 2v.

F. Saxl, 'Holbein's illustrations to the *Praise of Folly* of Erasmus', *Burlington Magazine*, LXXXIII (1943), 275, pl. I; Saxl and Meier, III, Taf. IV, 10–13, Taf. V, 14–17, Taf. XIV, 37, Tav. LV, 143; H. W. Janson, *Apes and Ape Lore in the Middle Ages and the Renaissance*, 1952, 113 n. 26, 214 n. 80, pls. XIIIb, XXXVIII; Pächt, Alexander, I, no. 886; T. Kurose, *Miniatures of the Goddess Fortune in Mediaeval Manuscripts*, 1977, pls. 157, 159.

1005. **Hertford College 3,** fols. 33–77 Saec. XV²
Scientific miscellanea. Latin translation by Garsia Arnaldi (?) de Suescun, Navarre, 1450 (241 × 171)

Good penwork initials. Diagrams. Humanistic script, fols. 37–43, probably by the translator, colophon fol. 37.

Prov. Bound up with earlier French MSS. with diagrams, fols. 1–36v, saec. XIV. Given to Magdalen Hall (Hertford College), 1613, by Matth. Freind of Westerby, Glos., fol. 1.

1006. **All Souls College 312** 1592
Valladolid
Patent of nobility granted to Juan de Prado Linares. In Spanish (300 × 224)

Good miniature, border. Written with calligraphic flourishes in a script close to that of Bodl. MSS. Buchanan c.2 and Bodley 909 (Pächt and Alexander, I, nos. 897, 898). [Illus.]

Contemporary Spanish binding of gold-tooled leather.

Prov. Arms of Philip II, first pastedown. Given by D. O. Malcolm, fellow, 1914.

ADDENDA TO FRENCH, FLEMISH, DUTCH AND ITALIAN SECTIONS

1007. **New College 369** *c.* 1500
Paris

Book of Hours (incomplete). Use of Amiens. In Latin and French (175 × 122)

Good miniatures (one defaced), borders, historiated initials. [Illus.]

Prov. Prayers written in a nearly contemporary hand by Claudine Faconnet. Calendar of Paris. Given by sir John Nicholson, 1983.

1008. **New College 372** Saec. XIII$^{3}_{4}$
Liège (?)

Psalter. Book of Hours (incomplete) (180 × 120)

Good historiated initials. Good penwork initials. [Illus.]

Prov. Sts. Gengulf, Begga, Foillanus in the litany. Collect for St. Lambert. Given by sir John Nicholson, 1983.

1009. **New College 370** Saec. XV med.
Flemish

Book of Hours. Use of Rome (160 × 105)

Good miniatures, borders. Good and other initials.

Prov. Given by Mrs. E. M. Vale, 1983.

1010. **New College 371** Saec. XV$^{3}_{4}$
Dutch

Book of Hours (incomplete). In Dutch (145 × 106)

Good historiated initial (defaced). Borders. Fine and other initials.

Contemporary (?) binding of brown stamped leather, rebacked.

Prov. Calendar of Utrecht. Notes in Dutch, end paste-down, saec. XVI. Members of the Nicholson family, 1898–1927. Given by sir John Nicholson, 1983.

1011. **New College 312,** fols. 39–231 1507
Dutch

Johannes de Monte Regio, Ephemerides. Misc. medical texts. In Latin and Dutch. On paper (210 × 145)

Tables, diagrams. Penwork initials. Written in 1507, fol. 221. Bound with various printed texts.

Prov. Given to William Daunton by Thomas Lewis of Oriel College, Oxford, fol. 1, saec. XVI.

1012. **Trinity College, Danson Collection 1**
Saec. XIV^{2}
North (?) Italy

Guillaume Peyraut, Summa virtutum (315 × 214)

Fine initial [Illus.]

Prov. Unidentified arms: argent, a fess gules. Erased *ex-libris*: 'Liber domini . . . de perusio decretorum doctoris quem emi a frater Abeldati (?)', saec. XIV, fol. 1. Erased *ex-libris* of the humanist Gasparino Barzizza (d. 1431), fol. 219^{v}. Bequeathed in 1445 to Sta Justina, Padua, no. 278 A. 13, by d. Antonio Zeno de Mediolano, vicar of the bp. of Padua, fol. 219^{v}. Bequeathed by John Raymond Danson, 1976.

L. Montobbio, 'Tre donazioni di codici a S. Giustina di Padova nel secolo XV°', *Studia patavina*, V (1958), 111, 121.

CONCORDANCE OF COLLEGE SHELF-MARKS

THE shelf-mark of each manuscript is given in heavy type followed by the Handlist number in parentheses

All Souls College:
1 (202), **2** (228), **4** (647), **6** (215), **7** (274), **10** (462), **11** (610), **12** (116), **14** (478), **15** (582), **18** (61), **21** (723), **26** (625), **27** (704), **29** (482), **30** (93), **36** (72), **37** (378), **40** (602), **42** (383), **44** (294), **49** (262), **50** (299), **51** (737), **52** (904), **55** (889), **57** (752), **59** (754), **68** (886), **71** (699), **72** (715), **73** (985), **74** (709), **79** (703), **83** (481), **85** (434), **86** (363), **88** (555), **89** (749), **92** (398), **94** (914), **95** (698), **98** (392), **302** (348), **312** (1006), **345** (523), **401** (91).

Balliol College:
1 (227), **2** (700, 895), **6** (37), **8** (669), **9** (98), **10** (119), **13** (68), **16** (504), **17** (505), **18** (507), **23** (115), **26** (557), **28** (443), **30** (511), **31** (847), **33** (327), **34** (577), **35A** (105), **35B** (928), **37** (554), **39–40** (195), **43** (696), **50** (421), **55** (855), **56** (747), **58** (858), **64** (743), **66** (861), **67A** (862), **67B** (496), **68** (497), **70** (498), **78A** (848), **78B** (946), **79** (853), **81** (592), **86** (356), **88** (563), **93** (408), **95** (321), **97** (566), **98** (499), **100** (500), **101** (325), **102** (902), **103** (706), **114** (270), **116** (250), **117** (567), **122** (438), **125** (856), **126** (935), **127** (491), **128** (939), **129** (257, 854), **130** (857), **131** (562), **132** (977), **135** (944), **137** (976), **138** (942), **139** (941), **140** (936), **148** (183), **150** (182), **151** (666), **154** (943), **156** (99), **158** (354), **166B,** **171** (384), **172** (188), **173A** (14), **173B** (104), **174** (526), **176** (114), **117** (180), **178A** (33), **178B** (118), **180** (852), **181** (860), **183** (874), **185** (968), **186** (454), **188** (97), **189** (573), **192** (414), **193** (667), **195** (688), **204** (579), **206** (580), **207** (512), **208** (728), **209** (613), **216** (503), **217** (386), **218** (49), **219** (460), **221** (272), **222** (211), **223** (73), **224A** (515), **224B** (859), **225** (355), **229** (128), **231** (720), **232A** (721), **232B** (284), **233** (945), **236** (514), **237** (938), **238A** (863), **238B** (864), **238C** (865), **238D** (866), **238E** (867, 1004), **240** (71), **242** (940), **245** (326), **248A** (930), **248B** (931), **248C** (932), **248D** (933), **248E** (934), **249** (937), **250** (680), **252** (251), **253** (673), **257** (109), **258** (576), **261** (506), **262** (929), **263** (424), **265** (849), **273** (9), **274** (845), **275** (718), **276** (442, 850), **277** (679), **279** (851), **280A–B** (192), **281** (893), **285** (652, 677), **286** (571), **287** (490), **288** (970), **290** (614), **291** (583), **295** (978), **296** (312), **300A** (517), **301** (289), **302** (727), **307** (13), **310** (447), **315** (444), **316A** (528), **316B** (485), **317** (54), **330** (247), **348** (221), **349** (529), **351** (193), **367** (875), **383** (789), **384** (807).

Brasenose College:
1 (676), **2** (238), **3** (187), **4** (229), **5** (181), **9** (430), **16** (626), **17** (624), **18** (876).

Campion Hall:
1 (873), **2** (783), **3** (816), **4** (998).

Christ Church:
87 (406), **88** (64), **92** (283), **93** (777), **94** (815), **95** (36), **98** (702), **100** (794), **101** (827), **105** (200), **106** (252), **107** (184), **109** (210), **111** (207), **113–14** (986), **115** (19), **121** (461), **145** (439), **146** (440), **148** (388), **150** (636), **151** (535), **178** (672), **179** (823), **180** (799), **507** (994).

Corpus Christi College:
1 (216), **2** (201), **2*** (25), **3** (185), **4** (407), **12** (394), **13–14** (825), **15** (120), **16** (92), **17** (371, 668), **18** (808), **31** (75), **50** (235), **52** (123), **56** (519), **59** (231), **61** (612), **65** (722), **66** (989), **67** (391), **69** (344), **70** (311), **73** (896), **78** (368), **79** (987), **82** (649), **84** (955), **88** (530), **89** (337), **91** (918), **95** (131, 236), **101** (869), **111** (697), **114** (166), **118** (35, 367), **122** (23), **123** (331), **124** (486), **129** (640), **144** (366), **147** (67), **153** (637), **157** (21), **159** (271), **161** (405), **162** (452), **189** (74), **193** (404), **198** (390), **201** (352), **207** (585), **209** (100), **218** (518), **219** (870), **224** (70), **236** (531), **241** (487), **255A** (882), **275** (253, 887), **282** (639), **285** (217), **350** (403), **385–6** (796), **387** (798), **394** (333), **410** (908).

Exeter College:
2 (382), **4** (8), **6** (360), **7** (402), **17** (324), **20** (675), **22** (314), **25** (110), **27** (520), **29** (281), **30** (288), **32** (204), **33** (232), **36** (451), **37** (426), **38** (150), **40** (136), **41** (401), **42** (279), **45** (178), **46** (306), **47** (316), **51–67** (586), **68** (593), **69** (705), **129** (596), **134** (585).

Hertford College:
2 (842), **3** (1005), **41** (868).

Jesus College:
32 (805), **37** (7), **38** (658), **40** (266), **43** (30), **50** (29), **52** (38), **53** (43), **55** (222), **56** (674), **59** (683), **62** (42), **63** (40), **64** (31), **65** (51), **66** (126), **67** (39), **69** (28), **70** (41), **91** (246), **93** (22), **102** (32), **105** (102), **107** (553), **108** (122), **110** (565), **114** (587), **124** (564).

Keble College:
1 (790), **2** (894), **3** (980), **5** (993), **6** (768), **7** (778), **8** (833), **9** (782), **10** (786), **11** (748), **12** (792), **13** (457), **14** (513), **15** (744), **16** (759), **17** (803), **18** (871), **19** (831), **20** (678), **21** (883), **23** (785), **26** (708), **27** (963), **28** (840), **29** (776), **30** (964), **31** (972), **32** (608), **33** (800), **34** (990), **35** (872), **38** (760), **39** (756), **40** (775), **41** (781), **42** (780), **43** (795), **44** (797), **45** (821), **46** (813), **47** (361), **49** (841), **53** (913), **54** (844), **55** (881), **56** (769), **57** (773), **58** (410), **59** (910), **60, 61, 62** (962), **63** (999), **65** (801), **66** (433), **67** (834), **68** (835), **69** (843), **76** (779), **77** (837), **78** (774), **79** (838), **80** (206), **81** (809), **Roll 1** (524).

Lady Margaret Hall:
Borough 1 (961, 997), **5** (561), **21** (635).

Lincoln College:
1 (736), **5** (295), **7** (256), **13** (108), **14** (757), **16** (174, 282), **17** (112), **19** (230), **20** (919), **21** (948), **23** (671), **24** (694), **26** (347), **27** (10, 125), **28** (521), **30** (111), **31** (197), **37** (949), **39** (950), **40** (973), **41** (983), **42** (951), **43** (979), **44** (959), **45** (958), **47** (957), **49** (952), **50** (735), **52** (459), **56** (522), **59** (953), **63** (69), **76** (901), **77** (591), **78** (380), **81** (501), **83** (137), **88** (489), **95** (892), **101** (441), **105** (369), **106** (417), **107** (341), **108** (377), **109** (719), **110** (379), **112** (455), **113** (282*bis*), **120** (784), **149** (812), **150** (692), **151** (456), **152** (770).

Magdalen College:
1 (681), **2** (244), **5** (758), **17** (107), **23** (516), **24** (165), **26** (15), **30** (425), **31** (46), **32** (329), **33** (411), **37** (926, 995), **39** (437), **41** (753), **44** (643), **47** (412), **50** (877), **52** (381), **59** (922), **62** (1000), **64** (984), **70** (59), **76** (965), **82** (488), **83** (982), **84** (78), **89** (607), **95** (502), **96** (550), **97** (416), **98** (336), **100** (171), **101** (213), **102** (63), **104** (375), **105** (60), **106** (163), **108** (189), **110** (376), **113** (878), **118** (113), **119** (77), **124** (167), **128** (103), **130** (168), **131** (169), **132** (170), **134** (364), **135** (923), **136** (924), **137** (365), **138** (556), **139** (127), **142** (905), **143** (464), **144** (559), **145** (458), **147** (525), **149** (172), **153** (466), **155** (285), **156** (446), **157** (536), **158–9** (538), **160** (467), **175** (663), **181** (338), **182** (969), **184** (975), **186** (484), **187** (717), **188** (539), **189** (463), **190** (476), **196** (575), **201** (606), **205** (330), **208** (374), **209** (300), **211** (594), **213** (600), **214** (599), **216** (891), **218** (474), **223** (828), **252** (817).

Mansfield College:
s.n. (996).

Merton College:
2 (243), **7** (190), **9** (142), **10** (340), **15** (139), **22** (301), **23** (618), **25** (619), **26** (541), **29** (159), **33–4** (296), **37** (298), **44** (724), **49** (537), **54** (335), **56** (263), **59–64** (568), **65** (572), **72** (218), **75** (713), **76** (712), **88** (80), **89** (588), **96** (261), **99** (730), **100** (729), **111** (248), **117** (399), **120** (191), **123–6** (716), **132** (657), **147** (313), **149** (734), **150–3** (584), **156** (343), **173–4** (598), **177** (95), **179** (534), **182** (160), **183** (138), **188** (273), **193** (475), **200** (650), **206–7** (687), **208** (161), **210** (175), **213** (239), **214** (346), **215** (373), **218** (653), **222** (690), **227** (733), **228** (906), **229** (903), **235** (242), **237** (686), **241** (642), **244** (292), **247** (493), **249** (164), **251** (413), **258** (310), **262** (309), **263** (880), **264** (897), **265** (483), **268** (574), **269** (241), **270** (701), **271** (684), **278** (255), **289** (277), **297B** (436), **298** (739), **300** (772, 900), **303** (766), **310** (351), **311** (76), **316** (79), **317** (802), **319** (472), **320** (291), **325** (177), **326** (960).

New College:
1 (665), **2** (225), **3–6** (695), **7** (173), **8–13** (533), **17** (662), **20** (205), **22** (711), **23** (302), **26** (682), **27** (196), **29** (198), **32** (83), **34** (117), **35** (96), **39** (477), **42** (630), **43** (629), **49** (449), **51** (278), **54** (27), **57** (631), **60** (879), **61** (628), **63** (927, 981), **65** (234), **66** (420), **87** (471), **89** (911), **90** (350), **101** (130), **104** (121), **105** (240), **107** (157), **113** (532), **114** (527), **115** (470), **116** (258), **124** (899), **129** (66), **130** (124), **134** (315), **144** (540), **152** (339), **154** (751), **155** (909), **157** (898), **159** (149), **167** (627), **174** (269, 670), **176** (888), **178** (344*bis*), **183** (308), **184** (907), **185–6** (755), **187** (249), **188** (287), **189–190** (750), **197–8** (746), **201** (921), **205** (885), **207** (290), **208** (286), **210** (133), **228** (569), **242** (323), **249** (925), **250** (966), **252** (645), **255** (1003), **256** (1001), **264** (276), **265** (912), **266** (492), **271** (609, 915), **274** (148), **277–9** (920), **280** (605), **281** (543), **283** (305), **284** (203), **286** (1002), **288** (581), **291** (570), **301** (47), **304** (832), **306** (216*bis*), **307** (615), **308** (106), **310** (544), **312** (1011), **313** (633), **315** (634), **316** (260), **318** (707), **319** (469), **322** (199), **323** (822), **326** (620), **342** (991), **358** (214), **363** (91), **369** (1007), **370** (1009), **371** (1010), **372** (1008).

Oriel College:
2 (18), **3** (3), **5** (545), **9** (226), **13** (297), **14** (233), **17** (319), **20** (427), **22** (884), **32** (445), **36** (94), **44** (732), **45** (320), **46** (280), **57** (322), **60** (151), **70** (435), **75** (400), **77** (176), **78** (224).

Pembroke College:
1 (468), **2** (358), **3** (152), **8** (656), **13** (659), **20** (846).

Queen's College:
52 (158), **55** (155), **70** (710), **168** (601), **207** (604), **210** (465, 830), **298** (791), **299** (693), **300** (132), **303** (154), **304** (431), **305** (788), **306** (81), **307** (357), **308** (328), **309** (62), **316** (162), **317** (153), **318** (651), **319** (26), **322** (156), **324** (396), **349** (818), **357** (622), **358** (223), **383** (372), **386** (648), **388** (397), **389** (303), **405** (767).

Radcliffe Trust
d. 60 (890).

Regent's Park College:
10.G (765).

St. Hilda's College:
s.n. (836).

St. John's College:
1 (268), **2** (359), **4** (220), **5** (134), **6** (595), **9** (661), **11** (44), **12** (793), **14** (345), **16** (254), **17** (5), **18** (495), **20** (24), **21** (147), **23** (590), **26, 27** (65), **28** (2, 4), **29** (219), **31** (129), **34** (654), **38** (45), **43** (101), **47** (453), **48** (186), **49** (89), **51** (144), **58** (510), **61** (146), **64** (829), **65** (318), **68** (811), **76** (307), **82** (814), **86** (370), **87** (611), **88** (16), **94** (418), **95** (82), **96** (84), **98** (731), **99** (48), **100** (179), **102** (761), **108** (293), **109** (429), **110** (208), **111** (90), **112** (143, 725), **113** (480), **115** (57), **116** (917), **117** (771), **123** (685), **125** (646), **128** (11), **129** (56), **130** (334), **131** (967), **132** (542), **150** (641), **152** (53), **164** (738), **165** (141), **167** (432), **172** (450), **173** (632), **174** (638), **178** (264), **179** (393), **187** (547), **192** (974), **193** (245), **194** (1, 508), **196** (597), **197** (762), **204** (726), **207** (691), **208** (763), **257** (621), **265** (819), **266** (820), **293** (494).

Somerville College:
1 (787), **2** (810).

Taylor Institution:
8° It. 2 (916), **8° It. 3** (947), **8° It. 4** (954).

Trinity College:
s.n. (473), **1** (558), **8** (332), **10** (617), **20** (644), **23** (415), **24** (209), **31** (34), **33** (50), **34** (86), **39** (20), **45** (55), **46** (589), **52** (58), **53** (664), **58** (88), **63** (52), **67** (689), **69** (87), **72** (135), **73** (623), **74** (145), **75** (578), **76** (603), **77** (362), **82** (194), **89** (839), **Danson Coll., 1** (1012)

University Archives:
A. 1 (275, 317), **B. 2** (616), **C. 3** (389).

University College:
5 (806), **7** (742), **8** (546), **9** (745), **12** (804), **17** (212), **37** (423), **40** (824), **54** (956), **56** (395), **61** (304), **62–3** (409), **71** (479), **78A** (422), **78B** (509), **82** (548), **85** (549), **86** (265), **87** (12), **94** (988), **100** (259), **112** (655), **115** (85), **116** (237), **120** (267), **129** (714), **130** (140), **154** (551), **165** (17), **168** (419), **169** (387), **177 A–B** (342), **178** (349), **181** (560).

Wadham College:
A. 7.8 (826), **1** (660), **2** (6), **5** (428), **10** (764), **15** (992).

Worcester College:
2.1 (353), **233** (448).

INDEX OF OTHER MANUSCRIPTS AND PRINTED BOOKS CITED

DATED OR DATABLE MANUSCRIPTS

British Manuscripts:

21 (1130–40), 38 (1147–76), 39 (1147–76), 40 (1147–76), 41 (1147–76), 42 (1147–76), 43 (1147–76), 64 (before 1167), 65 (*c.* 1170), 283 (1326–7), 316 (before 1373), 317 (1375), 331 (*c.* 1380), 332 (before 1388), 333 (1398), 386 (1402), 387 (1404), 388 (1405), 389 (1407–8), 418 (*c.* 1420), 425 (before 1417), 432 (1419–26), 434 (1427–8), 435 (1429), 436 (before 1435), 437 (*c.* 1437), 441 (*c.* 1440), 443 (1442), 445 (before 1444), 446 (1448), 447 (*c.* 1449), 490 (1450), 491 (1450), 495 (*c.* 1440–*c.* 1465), 496 (1445–53), 497 (1445–53), 498 (1445–53), 499 (1445–53), 500 (1445–53), 534 (1449), 562 (1445–53), 569 (1452), 570 (1452), 571 (*c.* 1454), 572 (1456), 574 (1458–9), 575 (1459 (?)), 576 (1461), 577 (1461), 578 (1461), 581 (*c.* 1463), 584 (before 1466–8), 585 (*c.* 1467–9), 586 (before 1469), 587 (1468–76), 589 (after 1471), 590 (*c.* 1473–5), 591 (before 1474), 592 (1475 (?)), 601 (after 1453), 602 (after 1453), 615 (1475–80), 616 (1477–8), 617 (*c.* 1484), 618 (1497), 619 (1497–1503), 633 (before 1536), 635 (1559), 636 (1561).

French Manuscripts:

660 (1244), 705 (1288), 725 (1308), 738 (1364–73), 754 (1406), 755 (1417), 757 (1415), 762 (1450), 772 (1449–80), 789 (1490–1503), 797 (1527), 799 (1599), 800 (*c.* 1600), 801 (1650).

Flemish Manuscripts:

823 (1503), 826 (1521), 827 (1528), 828 (*c.* 1529).

Dutch Manuscripts:

829 (1414), 832 (1468–9), 840 (1537), 1011 (1507).

German, Hungarian and Bohemian Manuscripts:

841 (*c.* 1267–76), 842 (1370–82), 845 (1409), 846 (*c.* 1410–19), 847 (1432), 848 (1442), 850 (1442–4), 856 (1442–4), 859 (1444), 860 (1444), 861 (*c.* 1444), 862 (*c.* 1444), 863 (*c.* 1444–5), 864 (1445–8), 865 (*c.* 1445), 866 (*c.* 1445), 867 (*c.* 1445), 869 (1472), 870 (*c.* 1484).

Italian Manuscripts:

909 (1379), 917 (1385 (?)), 925 (*c.* 1440–5), 926 (*c.* 1442), 927 (before 1444), 928 (*c.* 1444–53), 929 (*c.* 1444–53), 930 (*c.* 1444–53), 931 (*c.* 1444–53), 932 (*c.* 1444–53), 933 (*c.* 1444–53), 934 (1445), 935 (*c.* 1444–53), 936 (*c.* 1444–53), 937 (*c.* 1444–53), 938 (*c.* 1444–53), 939 (*c.* 1444–53), 940 (*c.* 1444–53), 941 (*c.* 1444–53), 942 (1444–53), 943 (1447), 944 (*c.* 1447–50), 945 (*c.* 1447–53), 946 (1448), 963 (1516), 964 (1404), 978 (1445), 979 (*c.* 1446), 980 (1447), 986 (1456), 989 (*c.* 1460), 992 (1523), 993 (1536), 995 (*c.* 1441), 997 (*c.* 1460–70), 999 (*c.* 1480), 1006 (1592).

INDEX OF TEXTS AND AUTHORS

THIS Index is of illustrated manuscripts only.

INDEX OF ILLUMINATORS AND SCRIBES

INDEX OF PERSONS, MAINLY OWNERS

COATS OF ARMS

(other than in Heraldic Treatises or Album amicorum)

MOTTOES

INDEX OF PLACES

LIST OF ILLUSTRATIONS

PLATES

PLATE I

1 ½

2 ⅓

3 ⅓

7 ½

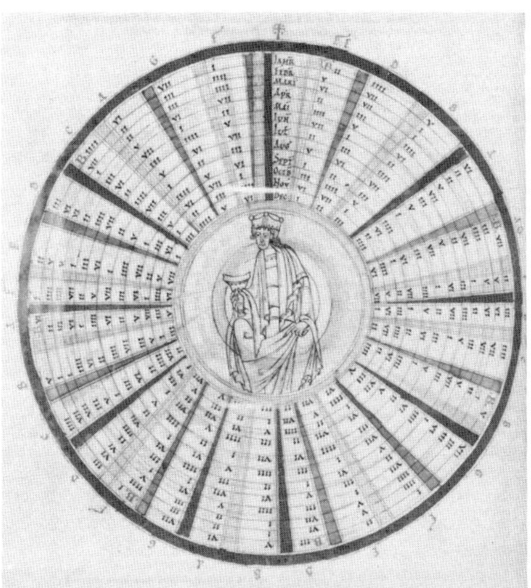

5 ⅓

6*a* ½

6*b* ½

8 ¾

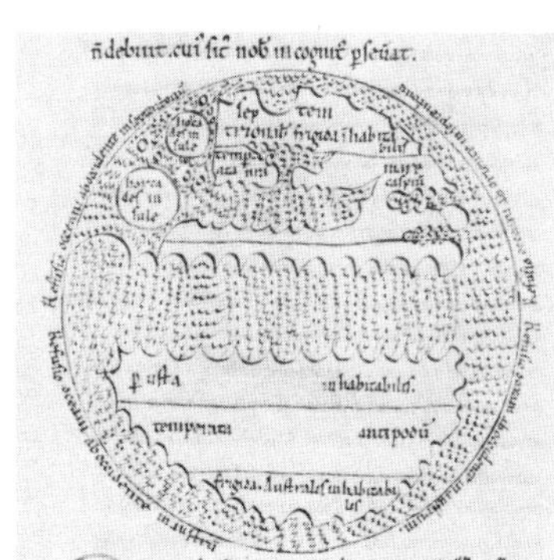

10 ½

13 ¾

PLATE II

11 $\frac{1}{1}$

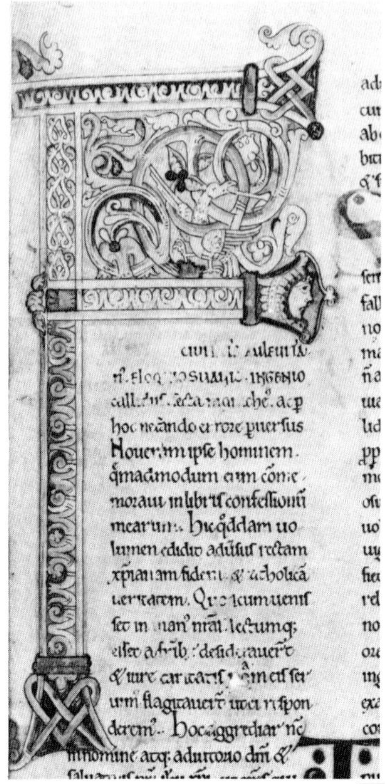

12 $\frac{1}{2}$

18 $\frac{1}{2}$

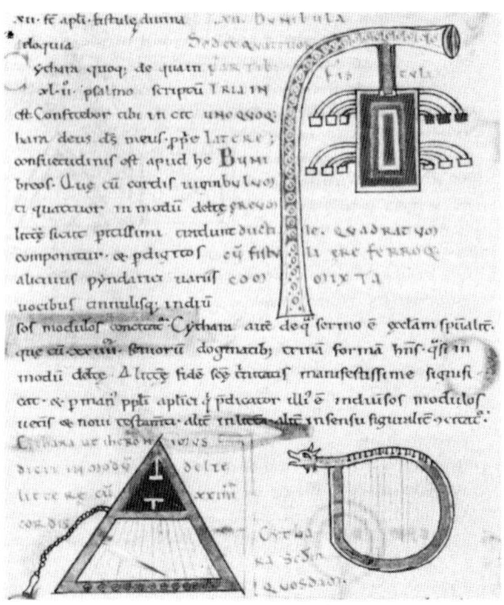

14 $\frac{1}{2}$

17*b* $\frac{3}{4}$

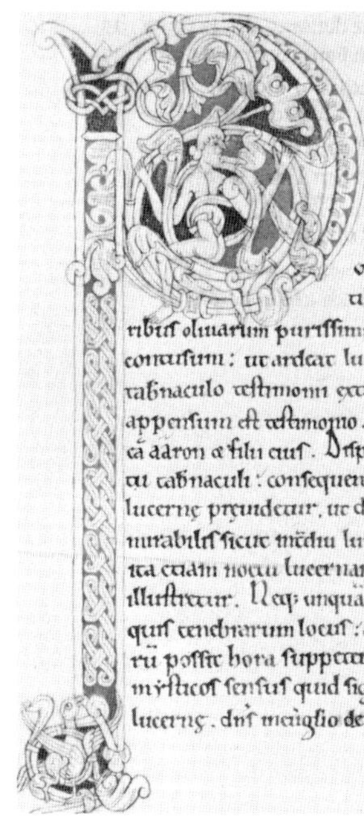

19 $\frac{1}{2}$

17*a* $\frac{3}{4}$

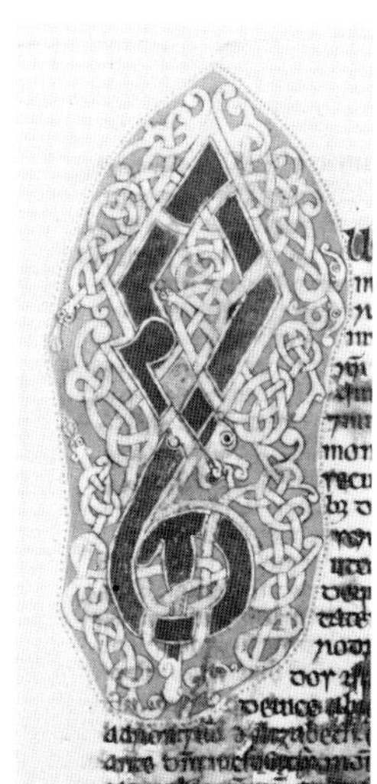

23 $\frac{1}{2}$

24 $\frac{3}{4}$

PLATE III

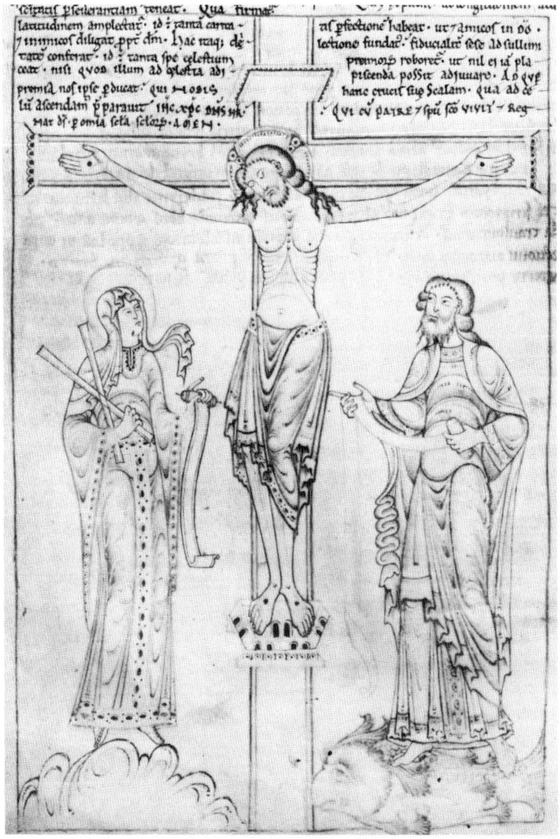

21 $\frac{1}{2}$

22 $\frac{1}{2}$

26 $\frac{1}{2}$

32 $\frac{1}{2}$

28 $\frac{1}{2}$

34 $\frac{3}{4}$

29 $\frac{1}{2}$

36 $\frac{3}{4}$

37a $\frac{3}{4}$

37b $\frac{1}{2}$

PLATE IV

25*a* 3/4

25*b* 3/4

40 1/2

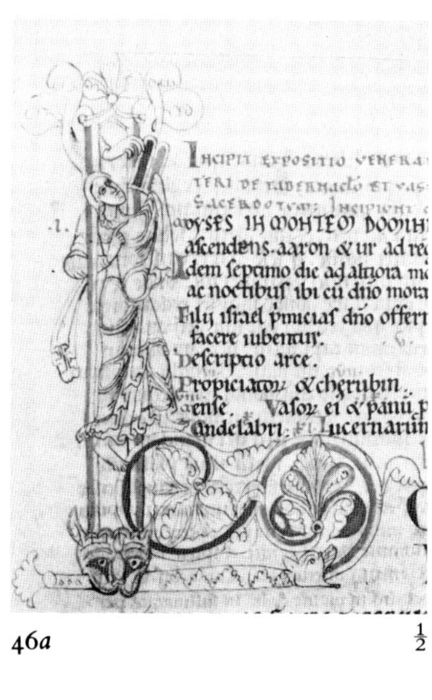

46*a* 1/2

46*b* 1/2

42 $\frac{1}{2}$

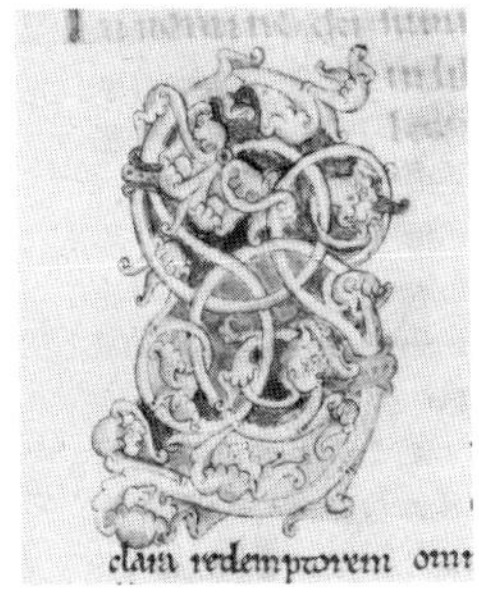
49 $\frac{1}{2}$

52 $\frac{1}{2}$

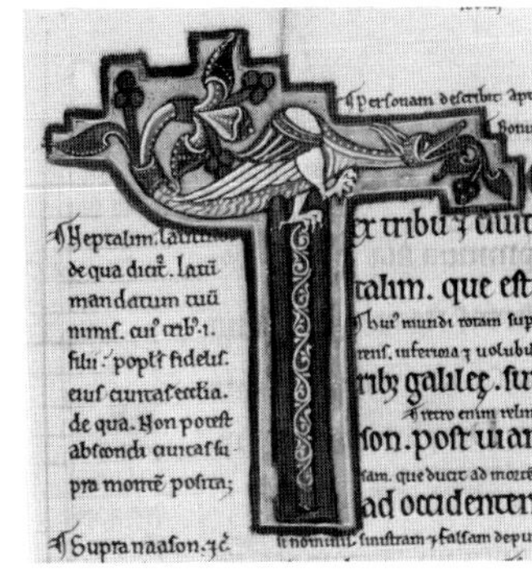
55 $\frac{1}{2}$

59 $\frac{3}{4}$

67 $\frac{3}{4}$

60 $\frac{3}{4}$

65 $\frac{3}{4}$

69 $\frac{3}{4}$

70 $\frac{1}{2}$

82 $\frac{1}{3}$

84 $\frac{2}{3}$

78 $\frac{1}{2}$

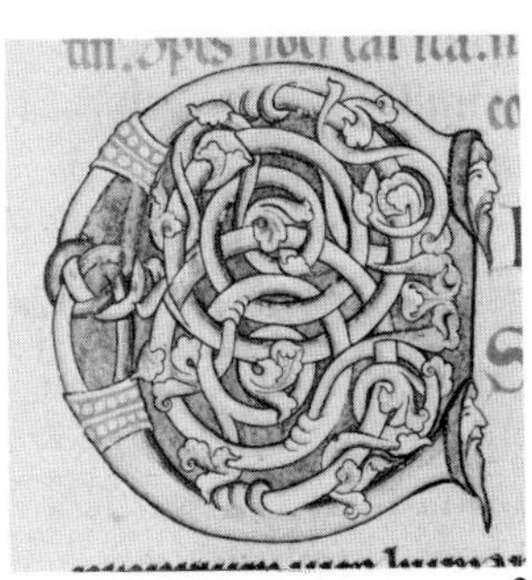
75 $\frac{2}{3}$

99 $\frac{1}{2}$

PLATE VI

88 $\frac{2}{3}$

89 $\frac{1}{1}$

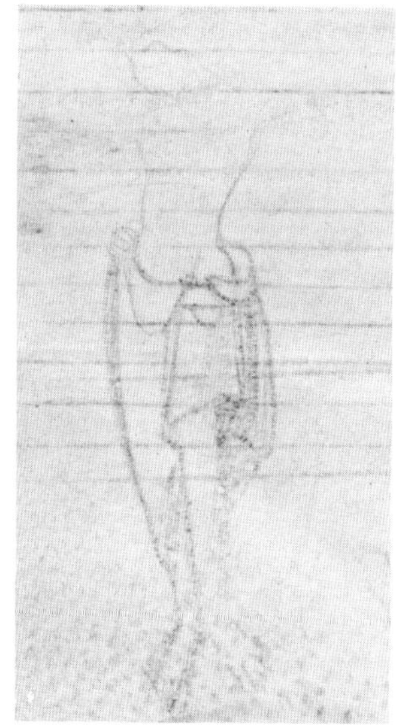

90 $\frac{1}{1}$

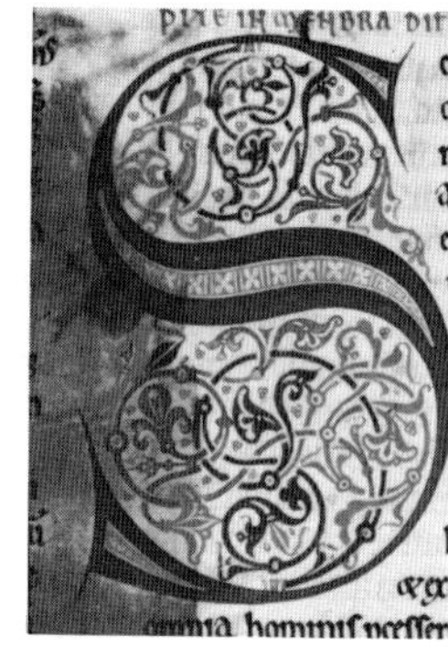

91 $\frac{1}{2}$

101 $\frac{3}{4}$

104 $\frac{2}{3}$

118 $\frac{1}{2}$

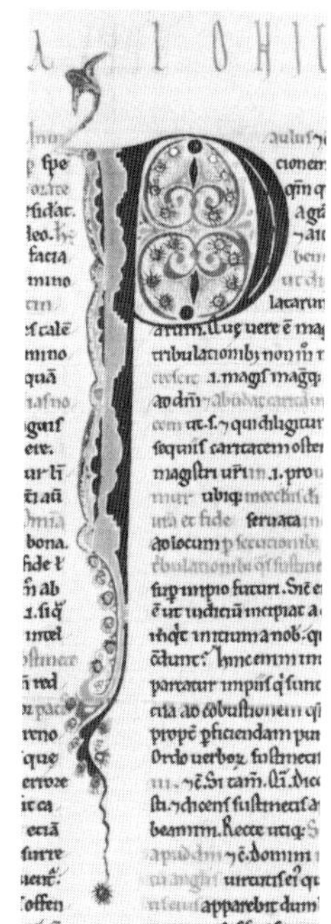

113 $\frac{1}{2}$

106*a* $\frac{3}{4}$

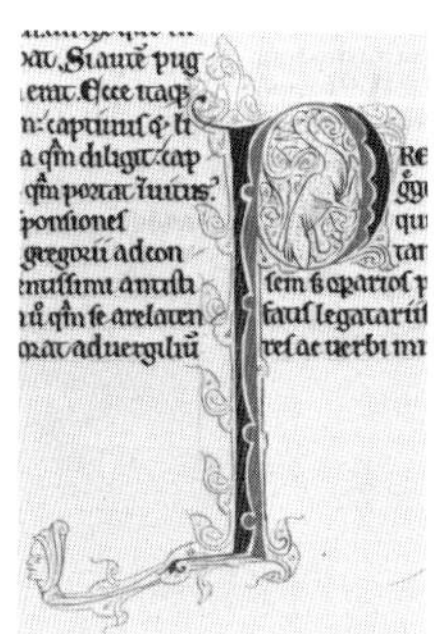

106*b* $\frac{1}{3}$

107*a* $\frac{2}{3}$

107*b* $\frac{1}{3}$

105 $\frac{1}{1}$

123 — 1/1

134*a* — 1/1

134*b* — 1/1

140 — 1/1

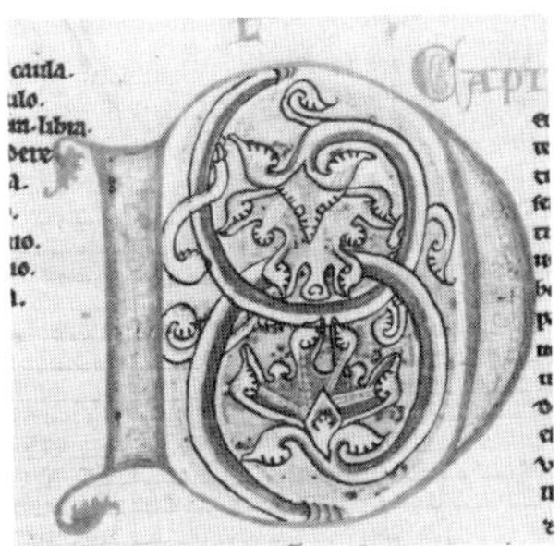
137 — 1/2

145 — 1/1

138 — 1/2

146*a* — 1/2

146*b* — 1/2

PLATE VIII

147 $\frac{2}{3}$

148 $\frac{2}{3}$

165 $\frac{1}{2}$

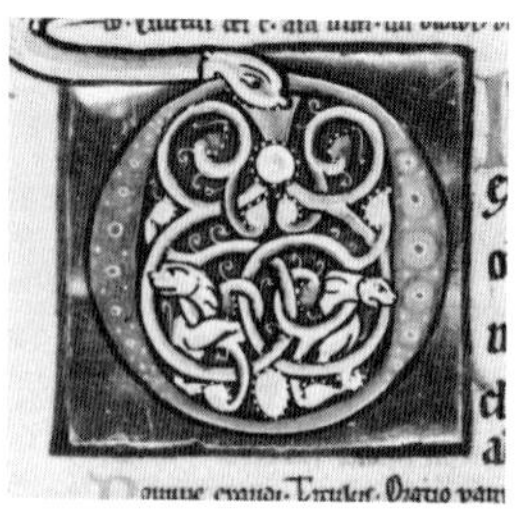

154 $\frac{2}{3}$

160 $\frac{2}{3}$

158 $\frac{2}{3}$

161*a* $\frac{1}{1}$

161*b* $\frac{1}{1}$

163 $\frac{3}{4}$

164 $\frac{3}{4}$

166 $\frac{1}{3}$

168 $\frac{1}{1}$

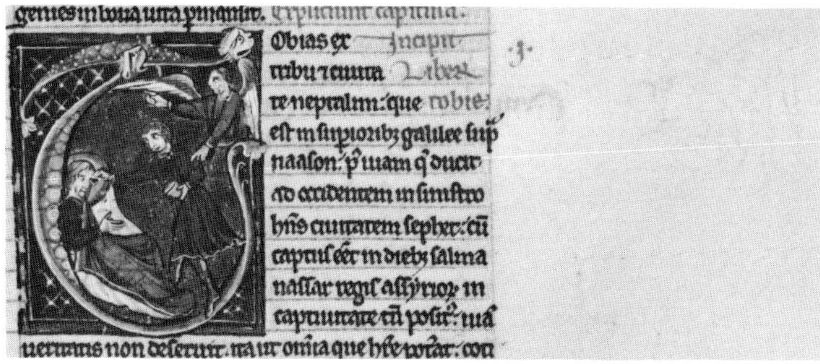

173*a* $\frac{1}{1}$

173*b* $\frac{1}{1}$

171*a* $\frac{2}{3}$

171*b* $\frac{1}{2}$

182 $\frac{1}{3}$

174 $\frac{1}{2}$

175 $\frac{1}{1}$

181 $\frac{1}{2}$

189 $\frac{1}{2}$

PLATE X

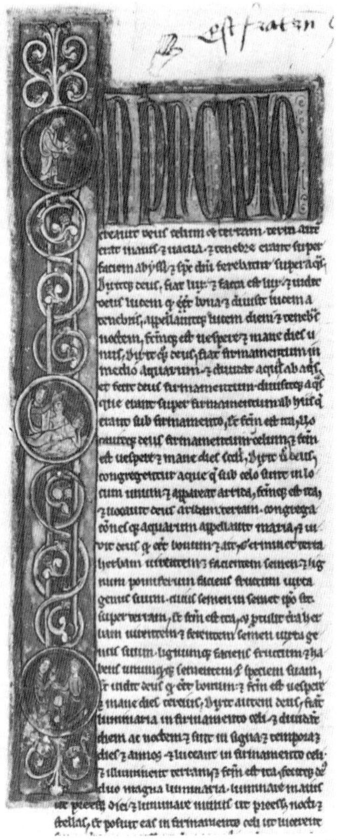

186 2/3

190*a* 3/4

190*b* 3/4

196 1/2

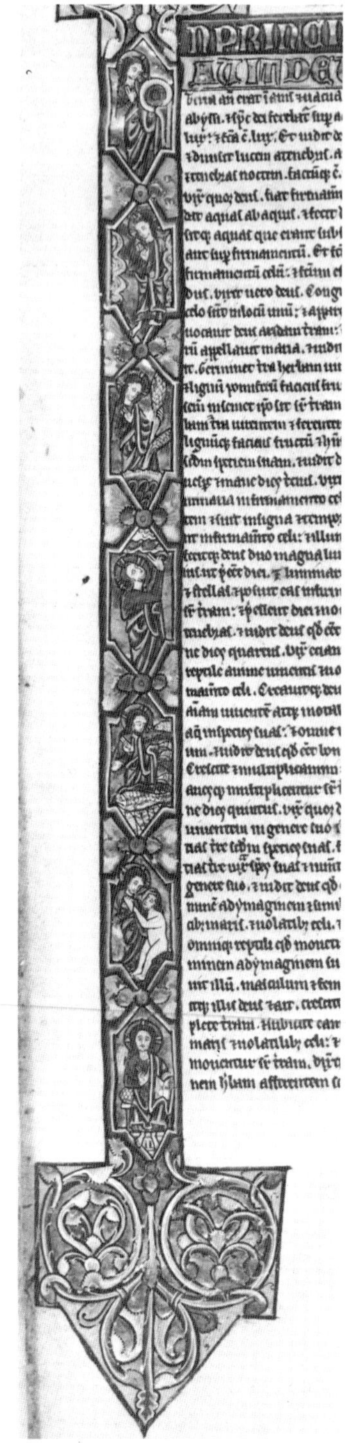

200 3/4

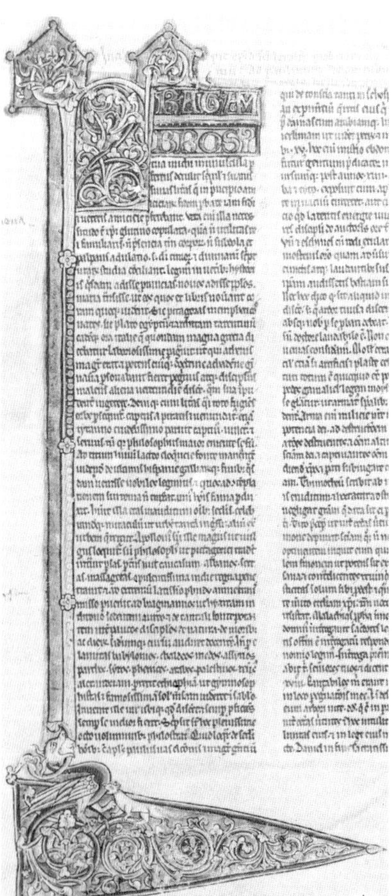

201 1/2

205*c* 3/4

205*b* 1/1

205*a* 1/1

199 $\frac{1}{1}$

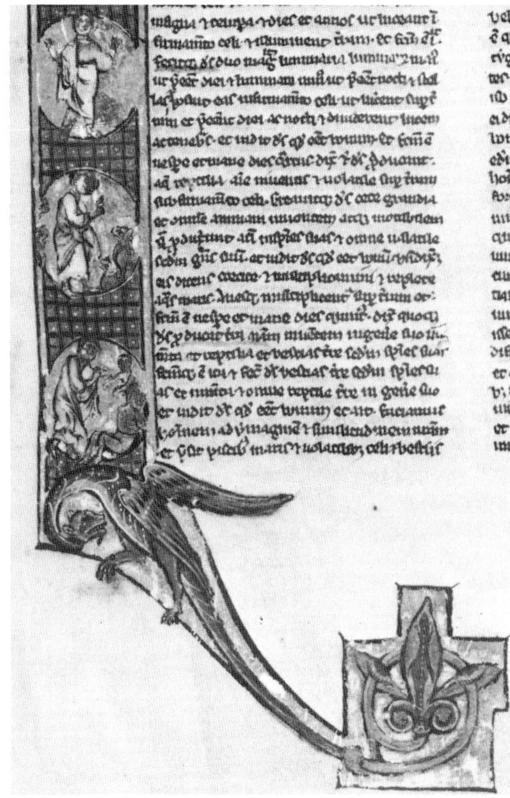

209 $\frac{1}{1}$

203 $\frac{1}{1}$

210 $\frac{1}{1}$

214 $\frac{1}{2}$

212 $\frac{1}{1}$

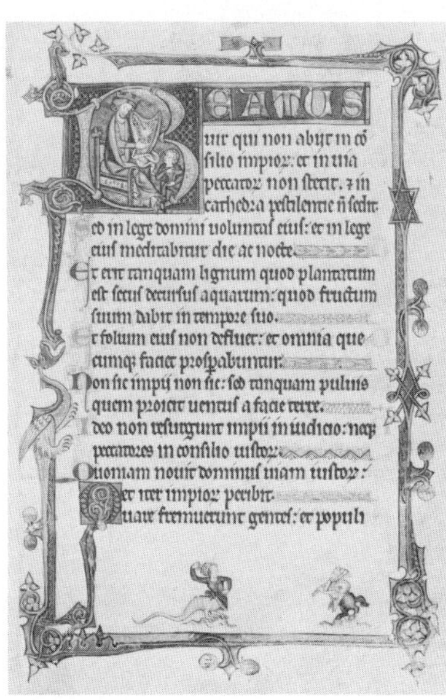

217 $\frac{1}{4}$

PLATE XII

215*a* $\frac{1}{2}$

215*b* $\frac{1}{2}$

219 $\frac{1}{2}$

215*c* $\frac{1}{3}$

215*d* $\frac{2}{3}$

220 $\frac{3}{4}$

218 $\frac{1}{1}$

216 *bis* $\frac{1}{1}$

216*a* $\frac{1}{1}$

216*b* $\frac{1}{1}$

216*c* $\frac{1}{1}$

223 $\frac{1}{2}$

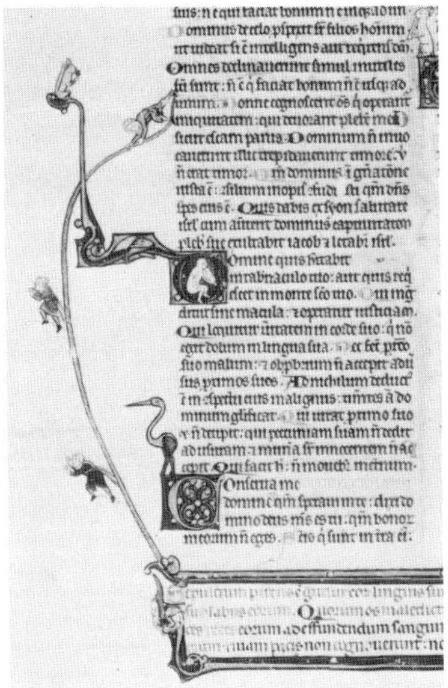

216*d* $\frac{1}{2}$

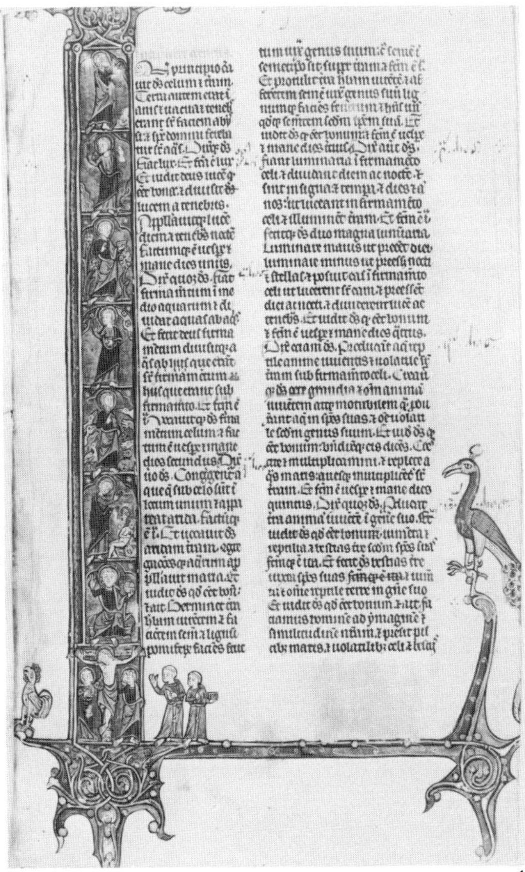

222*b* $\frac{1}{2}$

228*a* $\frac{1}{2}$

228*b* $\frac{1}{1}$

222*a* $\frac{1}{1}$

235 $\frac{1}{2}$

PLATE XIV

234 $\frac{3}{4}$

241*a* $\frac{1}{1}$

241*c* $\frac{1}{1}$

241*b* $\frac{1}{2}$

248 $\frac{3}{4}$

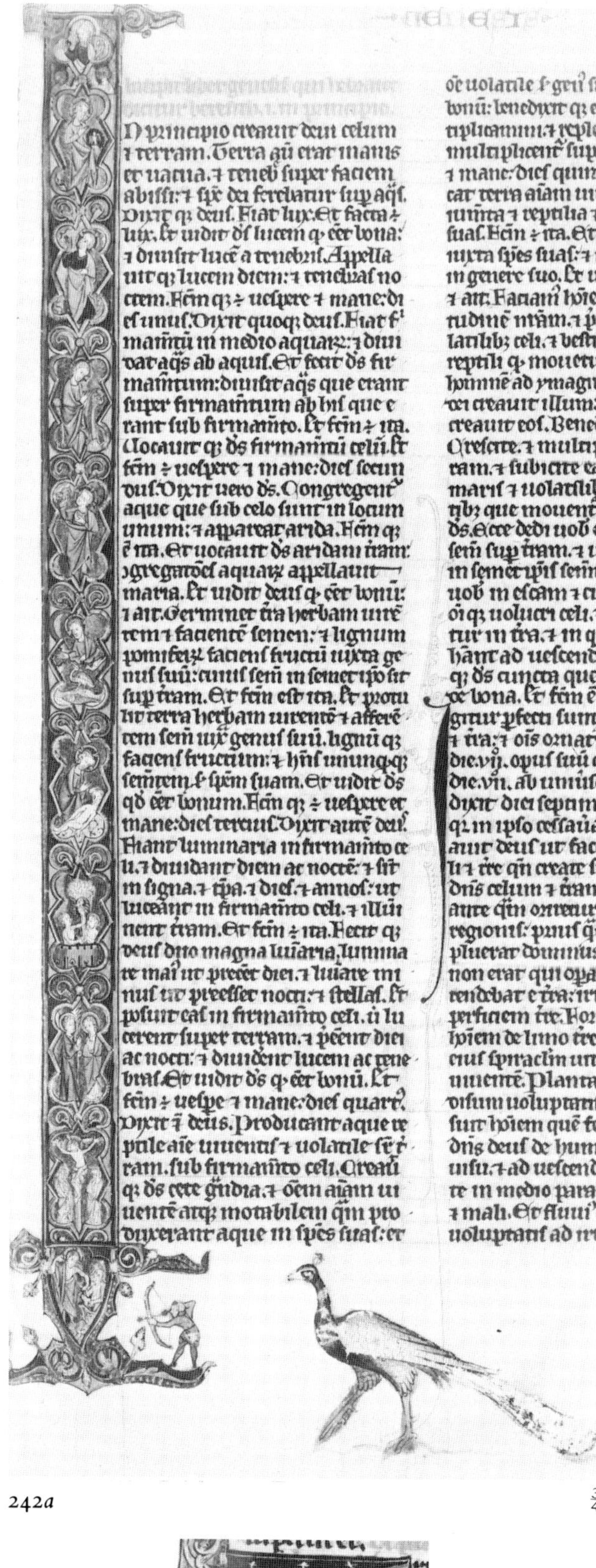

242*a* $\frac{3}{4}$

242*b* $\frac{1}{2}$

249a $\frac{7}{8}$

249b $\frac{1}{1}$

250 $\frac{3}{4}$

258 $\frac{1}{1}$

259b $\frac{1}{2}$

259a $\frac{1}{1}$

261 $\frac{1}{1}$

267 $\frac{1}{2}$

262 $\frac{2}{3}$

265 $\frac{3}{4}$

268 $\frac{1}{2}$

269 $\frac{3}{4}$

PLATE XVI

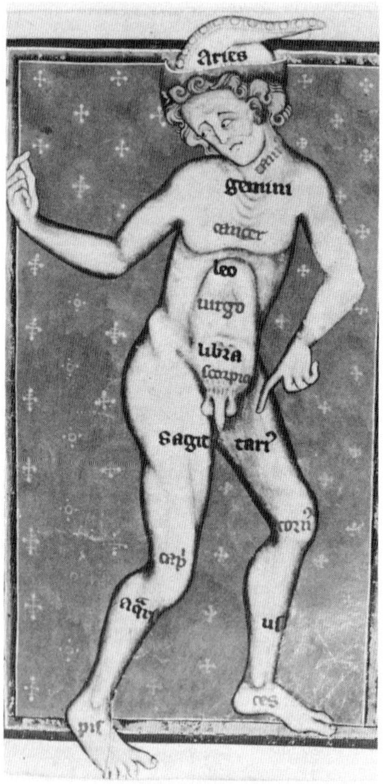

264*a* $\frac{1}{2}$

264*c* $\frac{1}{2}$

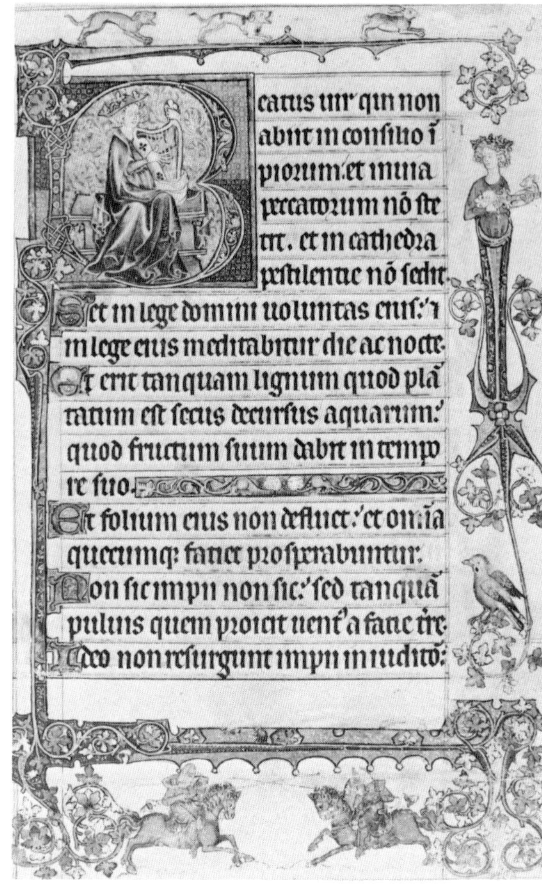

266*a* $\frac{1}{2}$

264*b* $\frac{2}{3}$

270 $\frac{1}{1}$

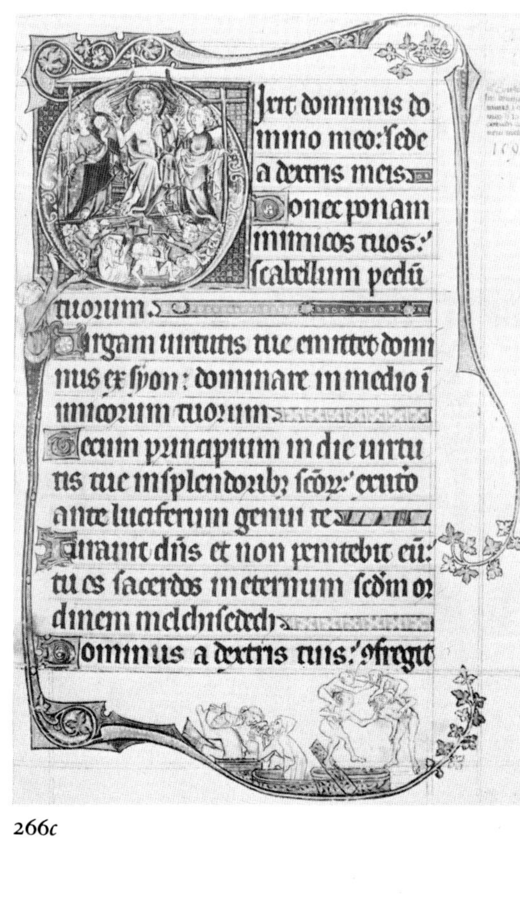

266*c* $\frac{1}{2}$

266*b* 1/1

274*a* 1/3

277*a* 1/1

274*b* 1/1

275 1/2

280*a* 3/4

277*b* 1/1

280*b* 1/1

279 $\frac{1}{1}$

282 *bis* $\frac{1}{1}$

282*a* $\frac{3}{5}$

282*b* $\frac{1}{1}$

283*a* $\frac{1}{2}$

283*b* $\frac{1}{1}$

283*d* $\frac{1}{2}$

283*e* $\frac{1}{2}$

PLATE XX

283*c* $\frac{1}{1}$

284*a* $\frac{3}{4}$

284*b* $\frac{1}{2}$

286*a* $\frac{3}{4}$

286*b* $\frac{1}{1}$

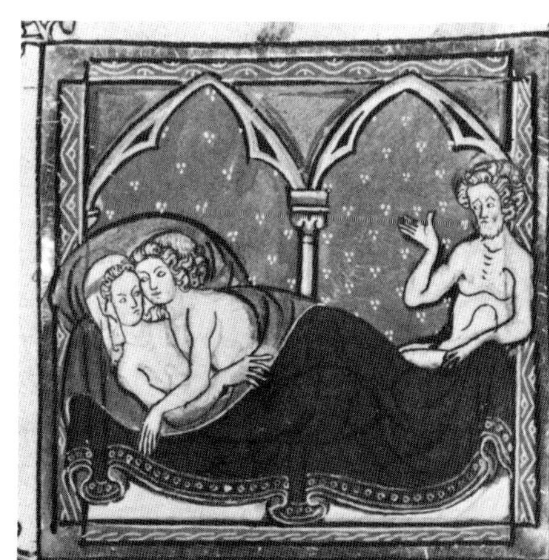

286*c* $\frac{1}{1}$

290 $\frac{3}{4}$

298 $\frac{1}{1}$

299 $\frac{3}{4}$

303 $\frac{1}{1}$

306*a* $\frac{3}{4}$

306*b* $\frac{1}{2}$

316*a* $\frac{3}{4}$

307 $\frac{1}{3}$

311*b* $\frac{1}{2}$

316*b* $\frac{2}{3}$

311*a* $\frac{1}{2}$

317 $\frac{2}{3}$

323 $\frac{2}{3}$

329 $\frac{3}{4}$

325 $\frac{2}{3}$

332*b* $\frac{1}{1}$

328 $\frac{1}{2}$

332*a* $\frac{1}{1}$

339 $\frac{1}{2}$

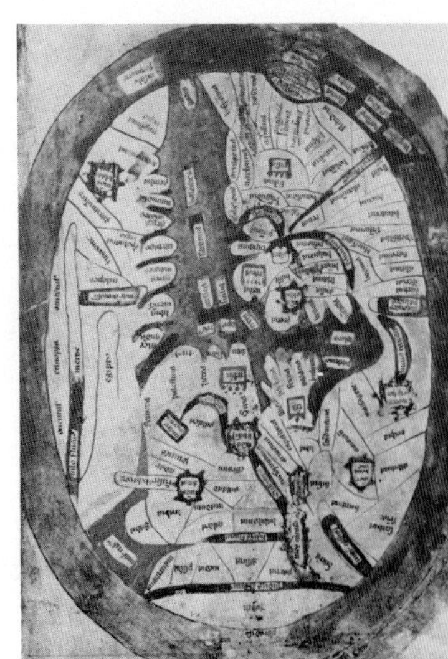

337 $\frac{1}{4}$

333 3/4

338 3/4

350 1/3

345 1/2

351 1/2

352 2/3

358 1/2

359 1/2

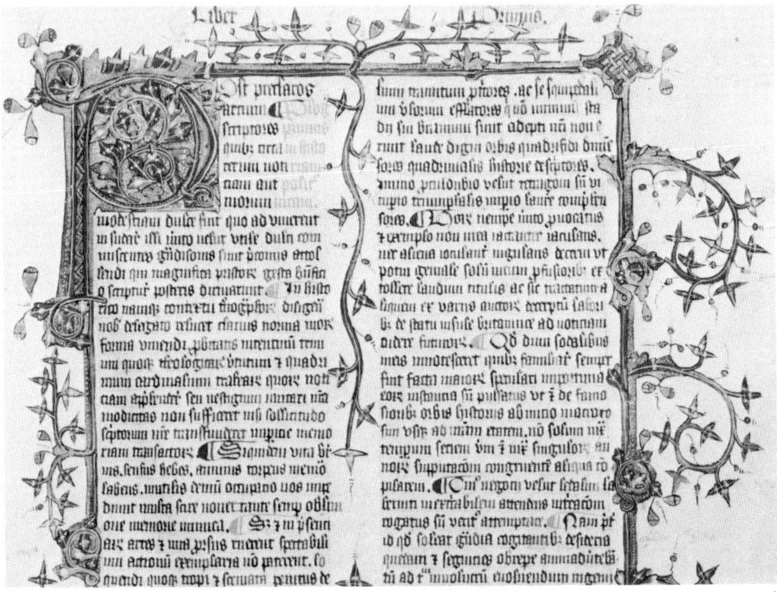

341 1/2

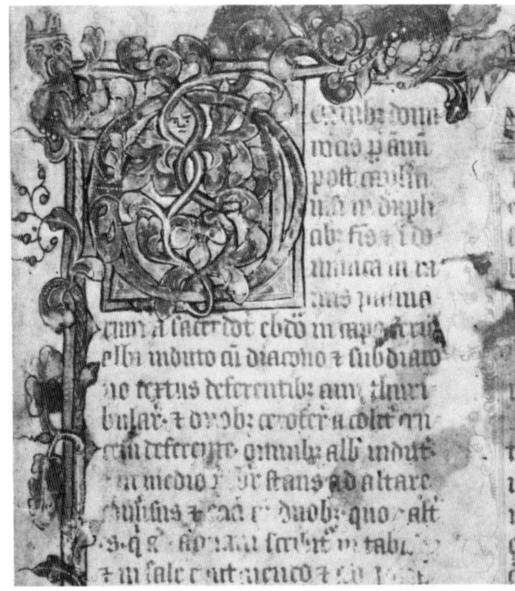

349 3/4

361*a* 1/1

361*c* 3/4

361*b* 3/4

PLATE XXVI

362 $\frac{1}{1}$

387 $\frac{1}{2}$

368 $\frac{3}{4}$

371 $\frac{3}{4}$

390 $\frac{1}{2}$

391 $\frac{1}{1}$

392 $\frac{1}{1}$

400 1/2

403 1/1

405*a* 1/2

405*b* 1/2

408 2/3

415 1/1

416 2/3

417 7/8

418 2/3

422 3/4

425 3/4

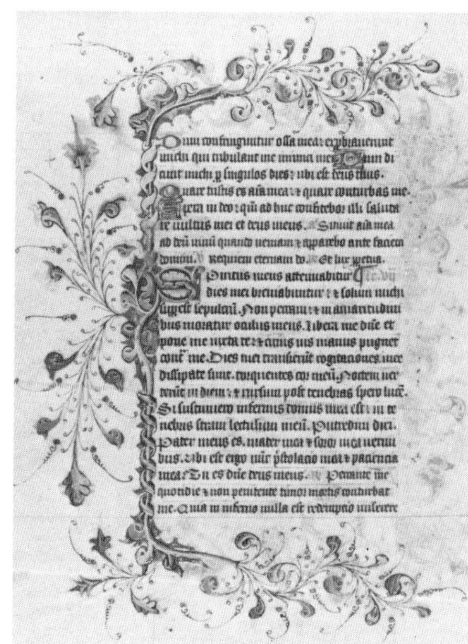

433 1/2

431*a* 3/4

431*b* 1/1

445 3/4

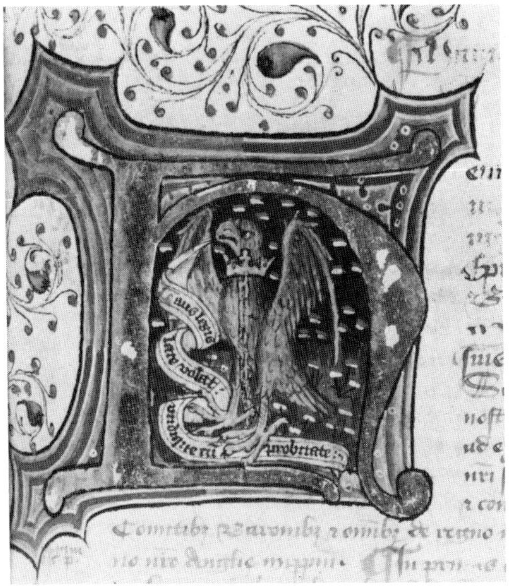

436*a* $\frac{3}{4}$

436*b* $\frac{3}{4}$

487 $\frac{1}{4}$

462 $\frac{1}{1}$

466 $\frac{3}{4}$

492 $\frac{1}{1}$

470 $\frac{1}{1}$

494 $\frac{3}{4}$

507 $\frac{2}{3}$

PLATE XXX

472*a* 3/4

495 1/2

472*b* 1/2

527*a* 2/3

518 3/4

509 3/4

527*b* 1/2

534 3/4

513*a* 2/3

516*a* 1/1

516*b* 2/3

526 2/3

556 1/2

513*b* 7/8

531 1/2

536 1/2

546 3/4

PLATE XXXII

548 $\frac{3}{4}$

570*a* $\frac{1}{2}$

570*b* $\frac{1}{2}$

549 $\frac{3}{4}$

564 $\frac{3}{4}$

568 $\frac{1}{2}$

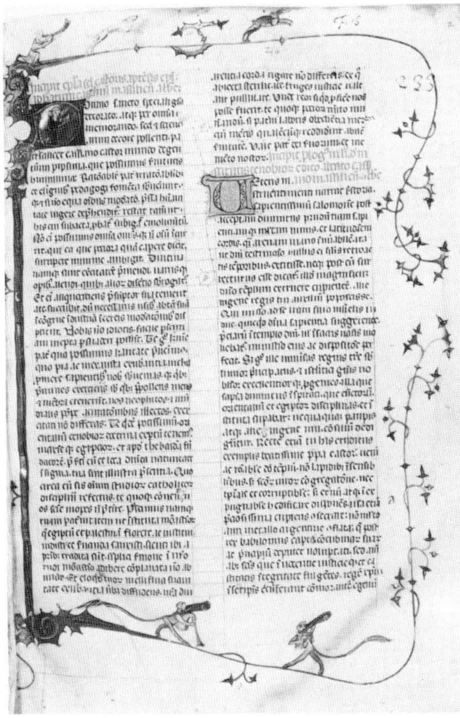

718 $\frac{1}{3}$

733 $\frac{3}{4}$

738*a* $\frac{1}{1}$

737 $\frac{2}{3}$

738*b* $\frac{1}{1}$

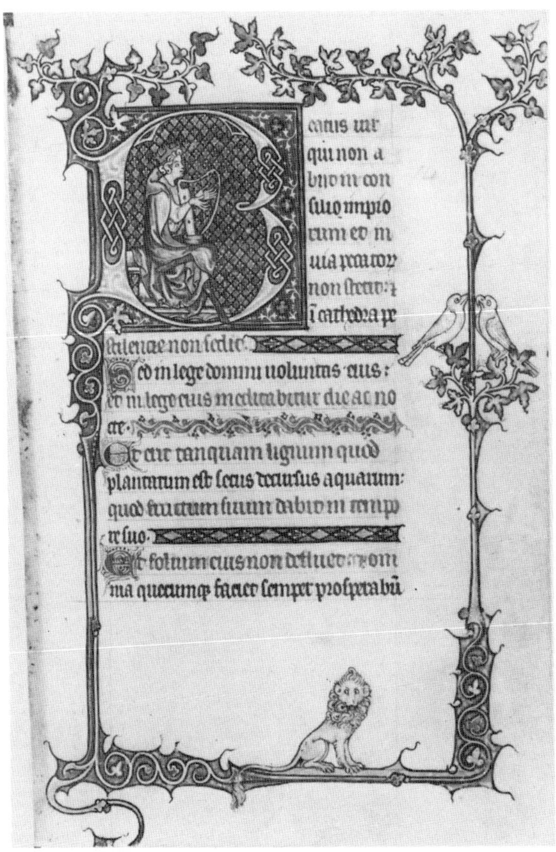

726 $\frac{3}{4}$

746 $\frac{1}{1}$

747 $\frac{1}{1}$

744a 1/1

748 3/4

744b 1/1

750 2/3

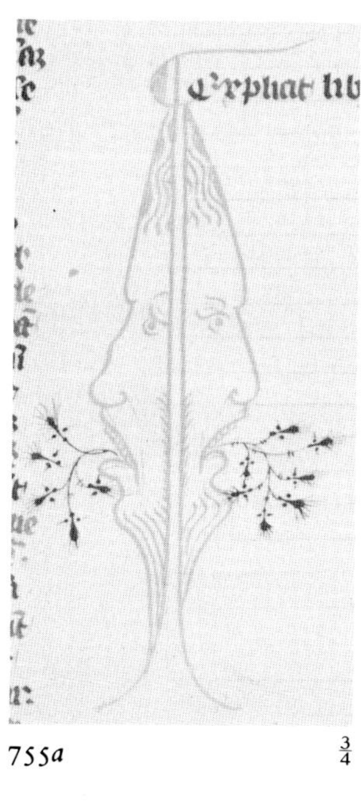

755a 3/4

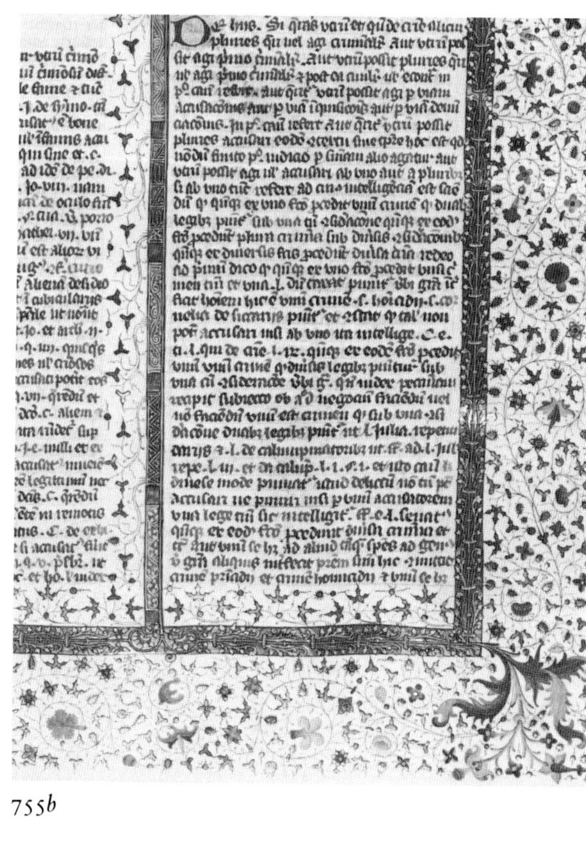

755b 1/2

758 $\frac{1}{1}$

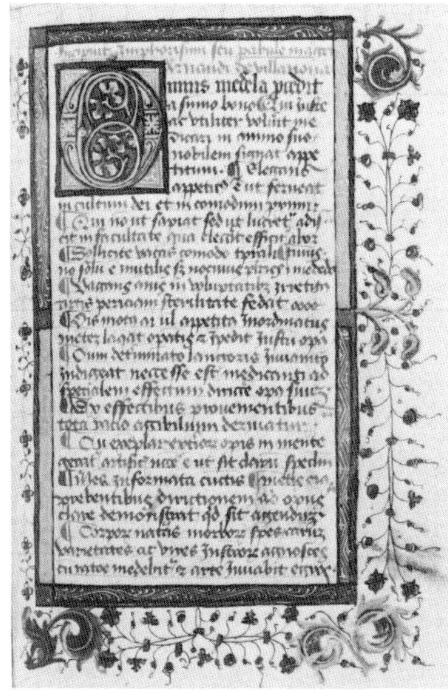

762 $\frac{1}{2}$

756*a* $\frac{3}{4}$

751 $\frac{3}{4}$

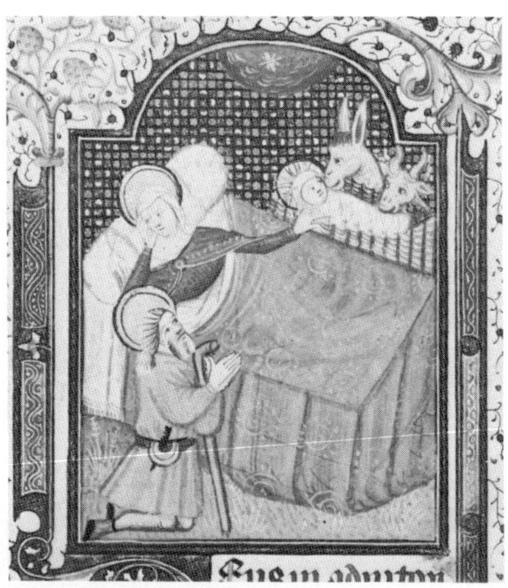

759 $\frac{2}{3}$

756*b* $\frac{2}{3}$

760 $\frac{1}{1}$

PLATE XLVI

763 $\frac{1}{1}$

770 $\frac{1}{1}$

769 $\frac{3}{5}$

764 $\frac{3}{4}$

765 $\frac{3}{4}$

767 $\frac{2}{3}$

773 $\frac{1}{2}$

775*a* $\frac{1}{1}$

775*b* $\frac{1}{1}$

791 $\frac{3}{4}$

792 $\frac{1}{2}$

PLATE XLVIII

777*a* $\frac{3}{4}$

777*b* $\frac{1}{1}$

788 $\frac{3}{4}$

781*a* $\frac{3}{4}$

781*b* $\frac{1}{2}$

778 $\frac{3}{4}$

780 $\frac{3}{4}$

789 $\frac{3}{4}$

785 $\frac{3}{4}$

794 $\frac{2}{3}$

797 $\frac{3}{4}$

PLATE L

796*a* ½

796*b* ½

795 ½

796*c* ¾

798 ½

800 ½

799 ½

803 ¾

804 1/1

PLATE LII

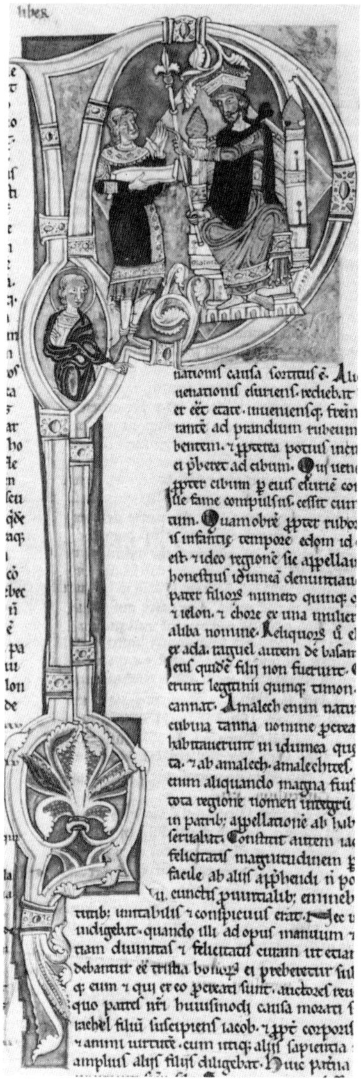

802*a* $\frac{1}{2}$

802*d* $\frac{3}{4}$

802*b* $\frac{1}{2}$

805*b* $\frac{3}{4}$

802*c* $\frac{1}{1}$

805*a* $\frac{3}{4}$

806 $\frac{1}{2}$

807 $\frac{1}{2}$

816 $\frac{1}{2}$

808 $\frac{5}{8}$

810 $\frac{1}{1}$

811 $\frac{1}{1}$

813 $\frac{1}{2}$

812 $\frac{1}{1}$

PLATE LIV

809 $\frac{2}{3}$

817 $\frac{1}{1}$

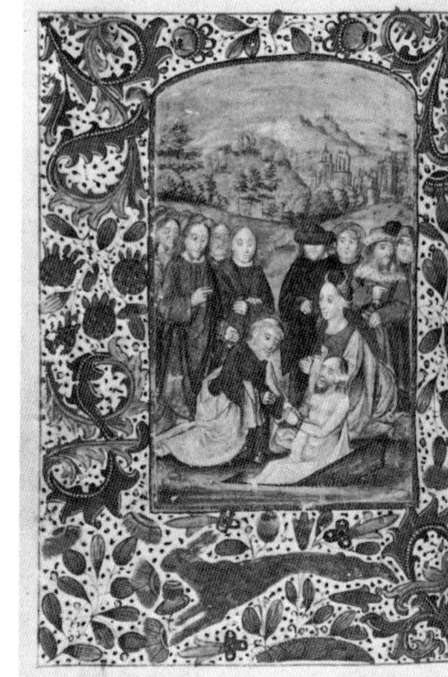

818 $\frac{1}{2}$

814*a* $\frac{1}{2}$

814*b* $\frac{1}{2}$

820 $\frac{3}{4}$

821 $\frac{1}{1}$

822*a* $\frac{2}{3}$

822*b* $\frac{2}{3}$

Homelia.xiii.diui Joannis Chrisostomi
ex euangelii sancti Matthei cap.v.& vi.
VDISTIS
quia di
ctum est
antiquis:
Diliges
proximu
tuu:& odio
habebis in
imicu tuu.
Ego aute
dico: Diligite inimicos ves tros:
Ecce sextum mandatum legem
adimplens: inimimu apud ho
mines odibiles, et inuicem odientes
qui inimicu odiri debere, improba
voce pronunciare non erubescunt.
Quod si iustum fuisset inimicos odire

824 $\frac{1}{2}$

PLATE LVI

823 3/5

826 1/2

825*b* 1/1

825*a* 2/3

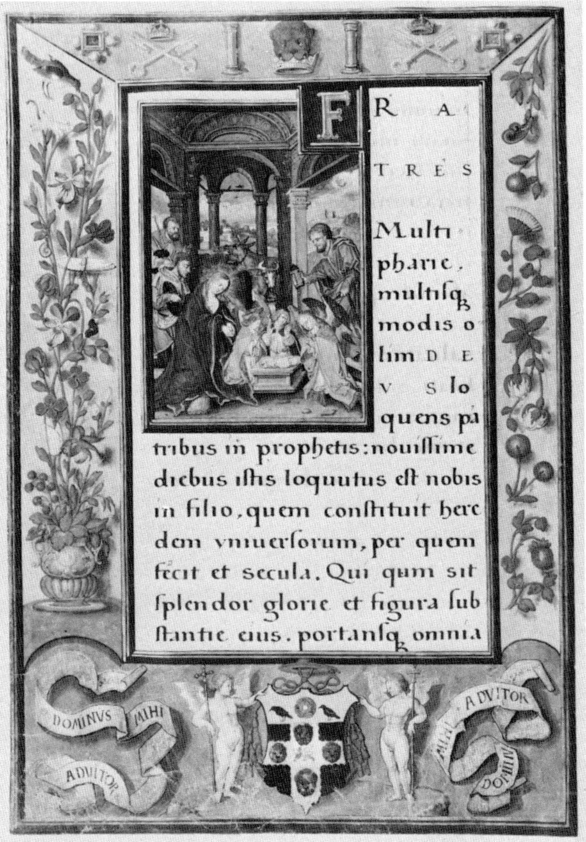

827*a* $\frac{2}{5}$

827*b* $\frac{2}{5}$

828*a* $\frac{3}{4}$

828*b* $\frac{1}{2}$

830 $\frac{1}{2}$

PLATE LVIII

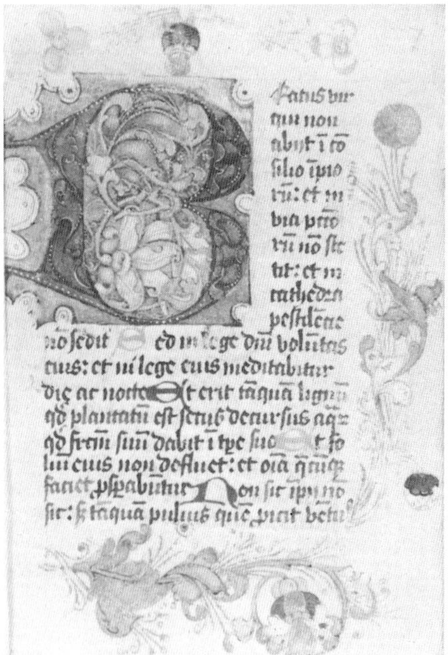

831 $\frac{3}{4}$

833 $\frac{2}{3}$

837 $\frac{1}{2}$

836*a* $\frac{3}{5}$

839 $\frac{1}{2}$

840 $\frac{3}{5}$

836*b* $\frac{1}{2}$

842 $\frac{1}{1}$

843 $\frac{7}{8}$

841*a* $\frac{1}{2}$

841*b* $\frac{1}{2}$

846 $\frac{3}{4}$

847 $\frac{2}{3}$

PLATE LX

844*a* 2/5

844*b* 1/2

853

873*a* 3/4

873*b* 2/3

872*a* $\frac{2}{3}$

872*e* $\frac{3}{5}$

872*b* $\frac{1}{2}$

872*c* $\frac{1}{2}$

872*d* $\frac{1}{2}$

PLATE LXII

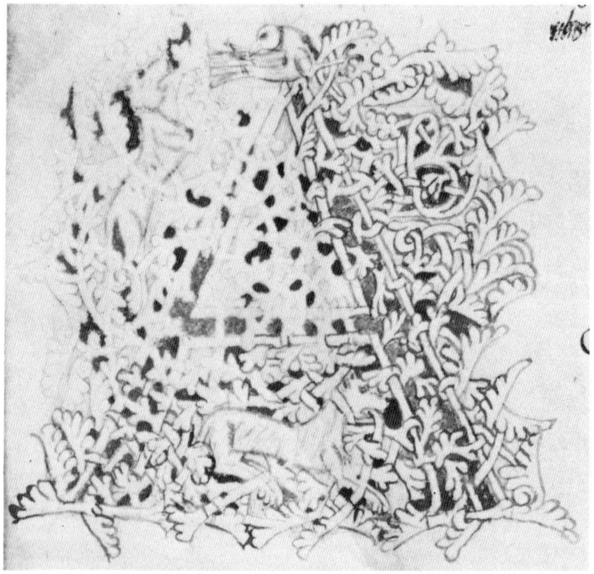

874 $\frac{2}{3}$

876 $\frac{1}{2}$

889 $\frac{1}{1}$

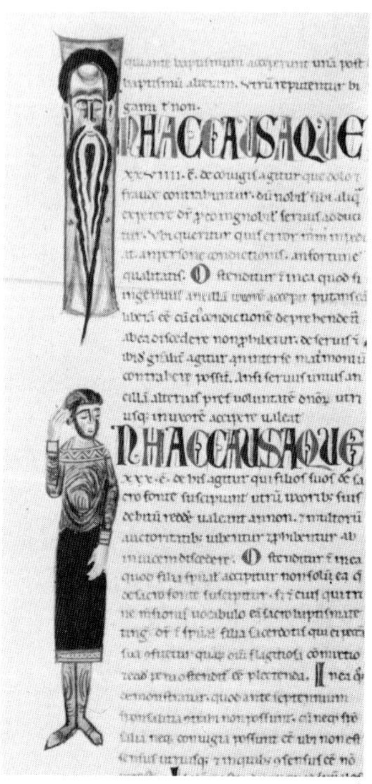

879 $\frac{1}{2}$

883 $\frac{1}{1}$

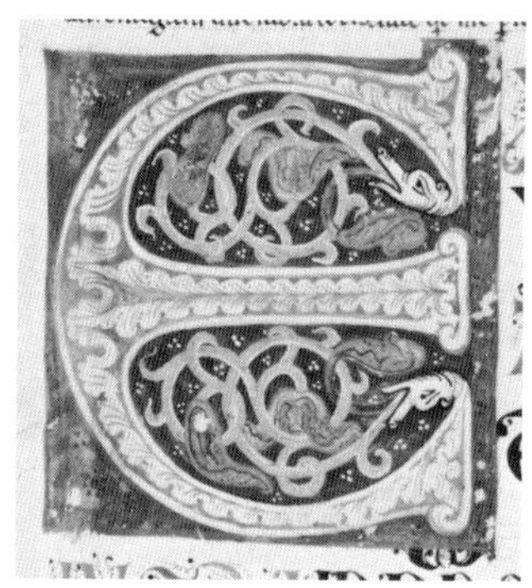

878*a* $\frac{2}{3}$

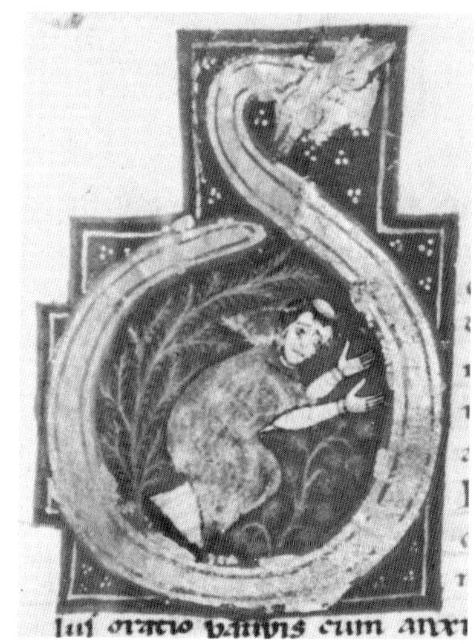

878*b* $\frac{1}{1}$

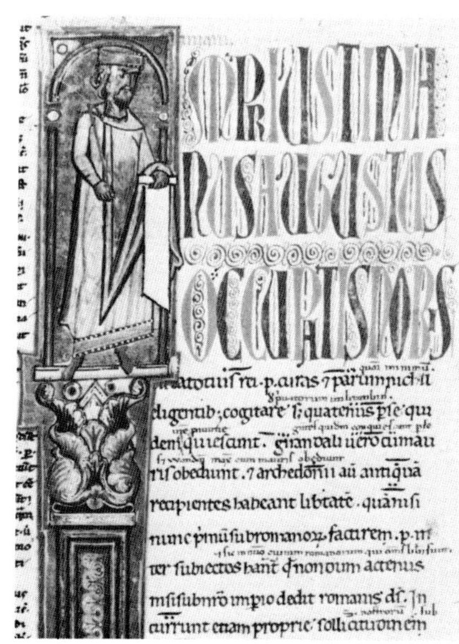

884 $\frac{3}{4}$

892 $\frac{3}{4}$

882 $\frac{1}{2}$

895 $\frac{3}{4}$

896 $\frac{3}{4}$

899 $\frac{3}{4}$

900*a* $\frac{1}{1}$

900*b* $\frac{1}{3}$

904 $\frac{2}{3}$

898 $\frac{3}{4}$

PLATE LXIV

903 $\frac{3}{4}$

908*a* $\frac{3}{4}$

908*b* $\frac{3}{4}$

909 $\frac{1}{1}$

910 $\frac{3}{4}$

917 $\frac{1}{1}$

907 $\frac{1}{1}$

912 $\frac{1}{2}$

914 $\frac{1}{4}$

924 $\frac{1}{2}$

920 $\frac{3}{4}$

923 $\frac{3}{4}$

927 $\frac{2}{3}$

925 $\frac{3}{4}$

931 $\frac{3}{4}$

936 $\frac{3}{4}$

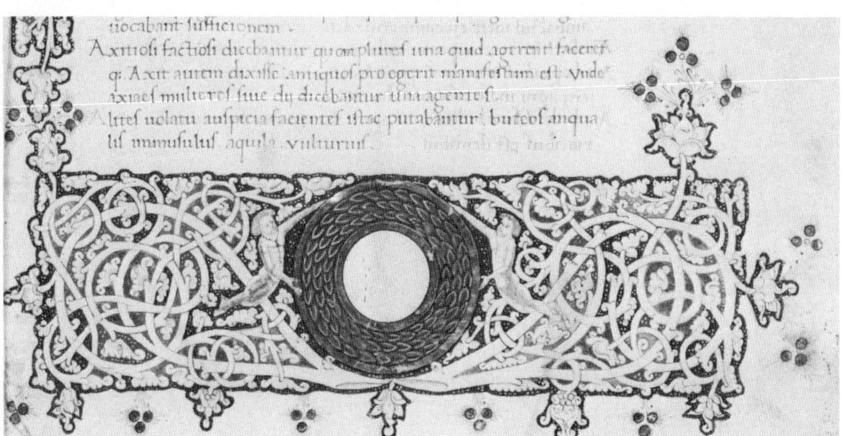

938 $\frac{1}{2}$

PLATE LXVI

937 $\frac{3}{4}$

943 $\frac{1}{1}$

946 $\frac{3}{4}$

950 $\frac{3}{4}$

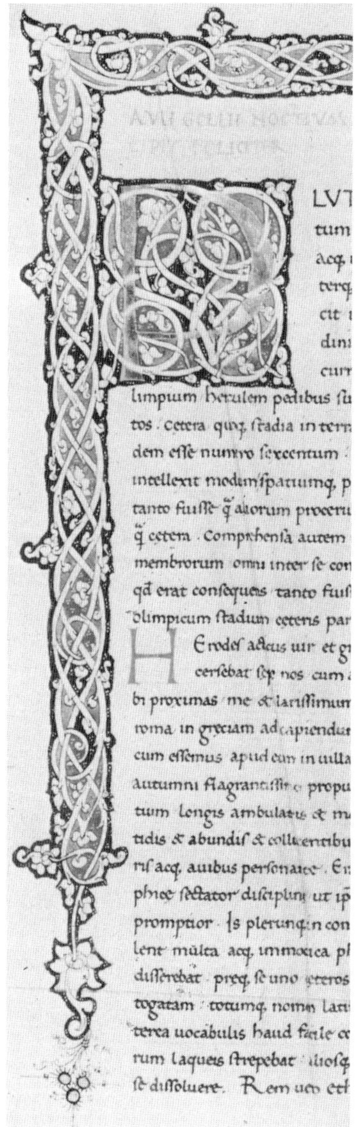

953 $\frac{2}{3}$

956 $\frac{1}{1}$

957 $\frac{3}{5}$

970 $\frac{7}{8}$

948 3/4

958 3/5

961 3/4

955 3/4

962 3/4

963 1/2

PLATE LXVIII

964*a* 1/2

967 4/5

964*b* 1/2

972 1/1

965 1/1

973 3/5

975 3/4

982 1/1

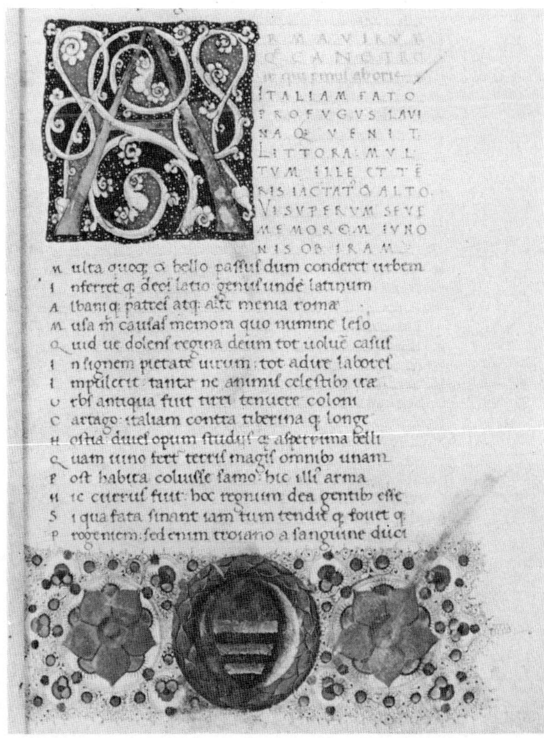
986 1/2

993 1/4

1003 1/2

989 1/1

991 1/1

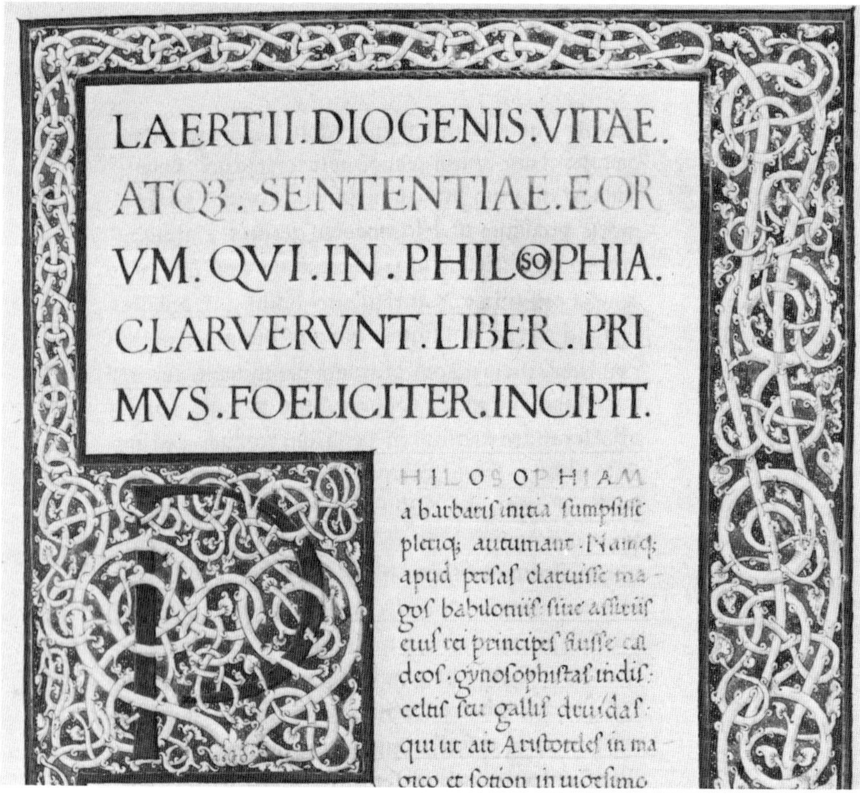

988 $\frac{3}{4}$

997 $\frac{1}{2}$

992 $\frac{1}{2}$

1004*a* $\frac{1}{2}$

1007 $\frac{1}{2}$

1004*b* $\frac{5}{8}$

999a

999b $\frac{3}{4}$

1006 $\frac{1}{3}$

1008 $\frac{3}{4}$

1012 $\frac{1}{3}$